Enzyme Technology

P. Gacesa and J. Hubble

Open University Press
Milton Keynes

Taylor & Francis
New York • Philadelphia

Co-published by:
Open University Press
Open University Educational Enterprises Limited
12 Cofferidge Close
Stony Stratford
Milton Keynes MK11 1BY, England

Taylor and Francis
Publishing office
3 East 44th Street
New York NY 10017
USA

Sales office
242 Cherry Street
Philadelphia, PA 19106
USA

First Published 1987

Copyright © 1987 P. Gacesa and J. Hubble

All rights reserved. No part of this work may be reproduced in any form, by mimeograph, or by any other means, without permission in writing from the publisher.

British Library Cataloguing in Publication Data
Gacesa, Peter
 Enzyme technology.—(Biotechnology series).
 1. Enzymes—Industrial applications
 I. Title II. Hubble, John III. Series
 661'.8   TP248.E5

ISBN 0-335-15168-x

ISBN 0-335-15160-4 Pbk

Library of Congress Cataloging in Publication Data
Gacesa, Peter
 Enzyme technology.
 Includes bibliographies and index.
 1. Enzymes—Industrial applications. 2. Biochemical engineering. I. Hubble, John. II. Title. [DNLM:
1. Biotechnology. 2. Enzymes. QU 135 G121e]
TP248.E5G33   1987   660'.63   87-10189

ISBN 0-8448-1515-2

ISBN 0-8448-1516-0 (pbk.)

Text design by Clarke Williams

Printed in Great Britain at the Alden Press, Oxford

Enzyme Technology

The Biotechnology Series

This series is designed to give undergraduates, graduates and practising scientists access to the many related disciplines in this fast developing area. It provides understanding both of the basic principles and of the industrial applications of biotechnology. By covering individual subjects in separate volumes a thorough and straightforward introduction to each field is provided for people of differing backgrounds.

Titles in the Series

Biotechnology: The Biological Principles M.D. Trevan, S. Boffey, K.H. Goulding and P. Stanbury
Fermentation Kinetics and Modelling: C.G. Sinclair and B. Kristiansen (Ed. J.D. Bu'lock)
Enzyme Technology: P. Gacesa and J. Hubble

Upcoming Titles

Monoclonal Antibodies
Biosensors
Industrial Fermentation
Plant Cell and Tissue Culture

Animal Cell Products
Chemical Engineering for Biotechnology
Waste Treatment
Bioreactors

Overall Series Editor

Professor J.F. Kennedy   Birmingham University, England

Series Editors

Professor J.A. Bryant   Exeter University, England
Dr R.N. Greenshields   Biotechnology Centre, Wales
Dr C.H. Self   Hammersmith Hospital, London, England

The Institute of Biology IoB

This series has been editorially approved by the **Institute of Biology** in London. The Institute is the professional body representing biologists in the UK. It sets standards, promotes education and training, conducts examinations, organizes local and national meetings, and publishes the journals **Biologist** and **Journal of Biological Education**.

For details about Institute membership write to: Institute of Biology, 20 Queensberry Place, London SW7 2DZ.

TP
248
.E5
G33
1987

To Karen, Tom, Luke and Sue

Contents

Preface — ix
Acknowledgements — xi
Symbols and Units — xiv

CHAPTER 1   **Introduction**   1
Historical perspective — 1
Choice of biocatalyst — 3
Legal implications in the use of enzymes — 6
Growth of the enzyme industry — 10

CHAPTER 2   **Commercial sources of enzymes**   14
Introduction — 14
Sources of enzymes — 14
Microbial enzymes — 16
Control of microbial enzyme production — 19
Genetic manipulation techniques — 22
Concluding remarks — 30

CHAPTER 3   **The extraction and purification of enzymes**   32
Introduction — 32
Enzyme extraction — 33
Enzyme purification — 35

Contents

	Large-scale purification	41
	Enzyme specification	43
	Concluding remarks	43

CHAPTER 4 Kinetic properties and reactor design   45
General considerations   45
Rate of reaction   45
Extent of reaction   53
Aspects of enzyme reactor design   55
Conclusions   64

CHAPTER 5 Medical and pharmaceutical applications of enzymes   65
Introduction   65
Enzyme therapy   66
Analytical uses   71
Pharmaceutical applications   73
Concluding remarks   76

CHAPTER 6 Effects of immobilization on enzyme stability and use   77
Introduction   77
Enzyme stability   77
Immobilization of enzymes   80
Conclusion   88

CHAPTER 7 Uses of enzymes in agriculture and the food industry   90
Introduction   90
Enhancement of traditional processes   91
Development of novel processes   94
Economic considerations   100

CHAPTER 8 Enzyme-based sensors   102
Introduction   102
Immobilized enzymes   103
Analytical reactors   104
Transducer-bound enzymes   105
Enzyme thermistors   109

	Enzyme field-effect transistors	115
	Direct enzyme–electrode interactions	116
	Other sensor devices	118
	Determination of biological oxygen demand	119
	Conclusions	120

CHAPTER 9 Approaches to enzyme modification 121

Introduction	121
Selection of the appropriate source of enzyme	121
Substitution of bound metal ions	122
Covalent modifications of enzymes	124
Enzymic modification of enzymes	125
Enzyme–coenzyme complexes	126
Non-specific mutagenesis	129
Site-specific mutagenesis	130

CHAPTER 10 Future prospects 137

Introduction	137
Prediction of enzyme folding/structure	138
Use of enzymes in organic solvents	141
Synthetic enzymes	144
Coenzyme regeneration	147
Concluding remarks	151

APPENDIX 1 Enzyme Commission nomenclature 152

APPENDIX 2 Residence time distribution analysis 159

The F diagram	159
Quantifying the deviation from an ideal state	160
Summary	161

APPENDIX 3 Design of enzyme assays 162

References 170
Index 177

Preface

Biotechnology has undoubtedly been one of the major growth areas in science and engineering over the last ten to fifteen years. The promise of new techniques with consequent development of novel processes has been publicized both in the scientific literature and in information that has been disseminated to a wider audience. Unfortunately, not all the claims for the potential of biotechnology have been based on sound analysis and the resultant over-selling of the topic has been a serious problem. However, the area of enzyme technology was not only well established before the current fervour for biotechnology but has grown successfully within it, providing a sound basis for a promising future.

In compiling this book we have aimed at producing a text that will be suitable for final year undergraduates, postgraduates, research workers and the technically-informed manager. The objective of the book is two-fold. We hope to give readers with an engineering background an appreciation of the subtleties of enzymes and the potential of the new techniques in molecular genetics for the tailoring of these catalysts to specific needs. For those with a biochemical/biological background who are more familiar with enzyme properties, we aim to provide an appreciation of biochemical engineering considerations. We do not claim to provide a comprehensive analysis of all the biochemical and engineering problems (there are several excellent texts already on the market) but rather to enable the interested reader to see an approach to a particular problem. Our philosophy has been to explain general principles, as far as this is possible, by using specific examples of enzyme applications. What we have tried to avoid is producing merely a catalogue of enzyme-catalysed processes.

It should be emphasized that from an engineering point of view enzymes are simply a special category of catalysts. They have several advantages in terms of specificity and mild reaction conditions but also disadvantages such as problems of instability. In some circumstances enzymes have replaced traditional catalysts or opened up new applications where, for example, specificity is a major criterion. However, our understanding of certain enzyme mechanisms has done much to further the development of conventional catalysis and chemists are now able to synthesize novel, low molecular weight, non-enzymic catalysts. Clearly, enzymes will not replace the majority of chemical catalysts and it may be argued that this application of enzymology is only a transient phase in the evolution of applied catalysis. However, what is clear is that the case for enzymes is well established and that the number of products of enzyme technology continues to increase. Also, the recent advances centred on the use of enzymes in non-aqueous media has the potential of opening up large new markets. It is likely that the application of enzymes to the production of new products will be the area of greatest growth potential rather than trying to cost-cut existing processes.

We hope that the book will widen the general awareness of the next generation of scientists and engineers to the commercial potential of enzymes, and will in some small part contribute to the development of this challenging field of enzyme technology.

Acknowledgements

We would like to express our thanks and appreciation for the invaluable advice received from a number of our colleagues including Dr R.A. John, Dr A.J. Knights, Professor J.A. Howell, and in particular to Dr Robert Eisenthal who was responsible for fostering our interest in enzymes during our period of study at Bath University.

We would also like to record our gratitude to Professor J.F. Richardson and the late Professor K.S. Dodgson who encouraged this venture and who supported us through the critical initial stages.

Chapter 1

Fig. 1.1   Lilly, M.D. (1977) *Biotechnological Applications of Proteins and Enzymes* (Bohak, Z. and Sharon, N. eds.) New York, Academic Press. (Original figure number 3, p. 135.)

Fig. 1.2   Solomons, G. (1977) *Biotechnological Applications of Proteins and Enzymes* (Bohak, Z. and Sharon, N. eds.) New York, Academic Press. (Original figure number 1, p. 52.)

Table 1.1   Lilly, M.D. (1977) *Biotechnological Applications of Proteins and Enzymes* (Bohak, Z. and Sharon, N. eds.) New York, Academic Press. (Original table number 4.)

Table 1.2   Reichelt, J. (1983) *Industrial Enzymology* (Godfrey, A. and Reichelt, J. eds.) Byfleet, The Nature Press. (Original table number 3.2.3., p. 149.)

Table 1.3   Reichelt, J. (1983) *Industrial Enzymology* (Godfrey, A. and Reichelt, J. eds.) Byfleet, The Nature Press. (Original table number 3.2.2., p. 148.)

Table 1.4   Aunstrup, K. (1977) *Biotechnological Application of Proteins and Enzymes*

(Bohak, K., and Sharon, N. eds.) New York, Academic Press. (Original table number 1, p. 40.)

Table 1.5   Poulsen, P.B. (1984) *Proceedings of the Third European Congress on Biotechnology*, Weinheim, V.C.H. (Original table number 4, p. IV–344.)

Chapter 2

Fig. 2.2   de Duve, C. (1985) *A Guided Tour of the Living Cell*, New York, Scientific American Books. (Diagram on page 93.)

Chapter 3

Fig. 3.1   Amicon Corporation, Lexington, MA. Copyright 1980.

Chapter 6

Fig. 6.3   Goldstein, L. (1976) *Methods in Enzymology* **44** (Mosbach, K. ed.) New York, Academic Press. (Original figure number 1, p. 403.)

Fig. 6.4   Horvath, C. and Engasser, J-M. (1974) *Biotechnology and Bioengineering* **16**, New York, John Wiley. (Original figure number 7, p. 919.)

Fig. 6.5   Engasser, J-M. and Horvath, C. (1976) *Applied Biochemistry and Bioengineering* **1**, New York, Academic Press. (Original figure number 6, p. 140.)

Chapter 7

Fig. 7.1   Antrium, R.L., Kolilla, W. and Schnyder, B.J. (1979) *Applied Biochemistry and Bioengineering* **2**, New York, Academic Press. (Original figure number 1, p. 126.)

Table 7.1   Godfrey, A. (1983) *Industrial Enzymology* (Godfrey, A. and Reichelt, J. eds.) Byfleet, The Nature Press. (Original table number 4.5.3, p. 227.)

Table 7.2   Godfrey, A. (1983) *Industrial Enzymology* (Godfrey, A. and Reichelt, J. eds.) Byfleet, The Nature Press. (Original table number 4.8.1, p. 295.)

Chapter 8

Fig. 8.5b   Mosbach, K. and Danielsson, B. (1981) *Analytical Chemistry* **53**, 83A–94A. (Original figure number 2.)

Fig. 8.7   Moss, S.D., Johnson, C.C. and Janata, J. (1978) *IEE Transactions in Biomedical Engineering* **25**, pp 49–54. (Original figure number 1.)

Acknowledgements

Fig. 8.8   Plotkin, E.V., Higgins, I.J. and Hill, H.A.O. (1981) *Biotechnology Letters* **3**, 187–192. (Original figure number 1.)

Table 8.1   Bowers, L.D. and Carr, P.W. (1980) *Advances in Biochemical Engineering* **15**, New York, Springer-Verlag. (Original table numbers 6 and 7, p. 106–107.)

Table 8.2   Mosbach, K. and Danielsson, B. (1981) *Analytical Chemistry* **53**, 83A-94A. (Original table number 1.)

Table 8.3   Lowe, C.R., Goldfinch, M.J. and Lias, R.J. (1984) *Biotech 83*, Northwood, Online Publications Ltd. (Original table number 1.)

Chapter 9

Fig. 9.2   Kaiser, E.T. and Lawrence, D.S. (1984) *Science* **226**, 505–511, Copyright 1984 by the AAAS. (Original figure 1.)

Table 9.2   Danno, G. (1970) *Agricultural and Biological Chemistry* **34**, 1805–1814. (Original table 1.)

Table 9.3   Jacobsen, H., Klenow, H. and Overgaard-Hansen, K. (1974). *European Journal of Biochemistry* **45**, 623–627. (Original table 1.)

Table 9.4   Kaiser, E.T. and Lawrence, D.S. (1984) *Science* **226**, 505–511, Copyright 1984 by the AAAS. (Original table 3.)

Chapter 10

Fig. 10.4   Zaks, A. and Klibanov, A.M. (1984) *Science* **224**, 1249–1251, Copyright 1984 by the AAAS. (Original figure 2A.)

Fig. 10.6   Bender, M.L., D'Souza, V.T. and Lu, X. (1986) *Trends in Biotechnology* **4**, 132–135, Copyright 1986 Elsevier Science Publishers. (Original figure 1.)

Symbols and units

A problem encountered in texts covering material which spans two or more traditional subject areas is that of constancy in the use of symbols and units.

Enzyme technology as presented in this text draws upon fundamental science arising from the study of enzymology, fluid dynamics and electronics. Each of these disciplines has its own well-established convention for use of symbols, making some overlap unavoidable. Rather than attempt to redefine all symbols to a common basis (and risk offending purists), we have defined them on the basis of the chapter in which they appear. For example, the symbol V denotes reactor volume in Chapter 4 whereas in Chapter 8 it denotes voltage.

In some disciplines it has been common practice to use various symbols to denote the same variable, e.g. the maximum rate for an enzyme reaction may be seen as V or V_m or V_{max}. While this diversity is usually frowned upon, it does allow us some scope to minimize conflict and to maintain the use of familiar symbols; to this end we must apologize for occasionally departing from use of the officially approved symbol.

In order to maintain consistency all dimensions are given in SI units or derived SI units of common usage. For a complete breakdown of derived SI units, the reader is referred to a specialist data book (Perry, 1984).

Although a rigid adherence to the SI nomenclature leads to problems with the magnitude of some terms (e.g. rate, $kg\ mol\ m^{-3}\ s^{-1}$, may lead to very small numbers in certain contexts), we feel that dimensional consistency must be stressed, especially for those readers from a biological background.

Symbols and units

Chapter 4

Symbol	Interpretation	Units
A	Arrhenius constant	depends on reaction order
D	Dilution rate	s^{-1}
ε	Packed bed voidage	—
E	Activation energy	$kJ\,kg\,mol^{-1}$
$[E]$	Active enzyme concentration	$kg\,mol\,m^{-3}$
$[ER]$	Concentration of enzyme–reactant complex	$kg\,mol\,m^{-3}$
$[E_0]$	Total active enzyme concentration	$kg\,mol\,m^{-3}$
$[E^t]$	Active enzyme concentration after time t	$kg\,mol\,m^{-3}$
$[ERR]$	Concentration of inactive enzyme–reactant complex	$kg\,mol\,m^{-3}$
$[EP]$	Concentration of inactive enzyme–product complex	$kg\,mol\,m^{-3}$
k_1	Second-order rate constant	$kg\,mol\,m^{-3}\,s^{-1}$
k_{-1}	First-order rate constant	s^{-1}
k_{-2}	Second-order rate constant	$kg\,mol\,m^{-3}\,s^{-1}$
k_2	First-order rate constant	s^{-1}
k_d	First-order decay constant	s^{-1}
K_m	Michaelis constant	$kg\,mol\,m^{-3}$
$K_{\dot{m}}$	Apparent Michaelis constant	$kg\,mol\,m^{-3}$
K_i	Inhibition constant	$kg\,mol\,m^{-3}$
K_{eq}	Equilibrium constant	—
$[P]$	Product concentration	$kg\,mol\,m^{-3}$
Q	Volumetric flow rate	$m^3\,s^{-1}$
R	Gas constant	$kJ\,K^{-1}\,kg\,mol^{-1}$
$[R]$	Reactant (i.e. substrate) concentration	$kg\,mol\,m^{-3}$
T	Temperature	K
v	Observed rate of reaction	$kg\,mol\,m^{-3}\,s^{-1}$
V	Reactor volume	m^3
V_l	Liquid volume	m^3
V_{tot}	Total volume	m^3
V_{max}	Maximum theoretical rate of reaction	$kg\,mol\,m^{-3}\,s^{-1}$
$V_{ma\dot{x}}$	Apparent V_{max}	$kg\,mol\,m^{-3}\,s^{-1}$
X	Fractional conversion	—

Chapter 6

Symbol	Interpretation	Units
$a,b,c,$	Constants	—
C_b	Bulk concentration	$kg\,mol\,m^{-3}$
C_s	Surface concentration	$kg\,mol\,m^{-3}$
d_i	Diameter of impeller	m

Symbol	Interpretation	Units
d_p	Diameter of particle	m
D_c	Diffusivity of solute through the immobilization matrix	$m^2 s^{-1}$
D_s	Diffusivity of solute in solution	$m^2 s^{-1}$
e	Electronic charge	—
E	Active enzyme concentration	$kg\,mol\,m^{-3}$
E_o	Active enzyme concentration at time zero	$kg\,mol\,m^{-3}$
E^t	Active enzyme concentration after time t	$kg\,mol\,m^{-3}$
H_b^+	Hydrogen ion concentration in bulk solution	$kg\,mol\,m^{-3}$
H_s^+	Hydrogen ion concentration at the surface	$kg\,mol\,m^{-3}$
k	Boltzmann constant	$J\,K^{-1}$
k_d	Enzyme decay rate	s^{-1}
K_m'	Apparent Michaelis constant	$kg\,mol\,m^{-3}$
K_s	Mass transfer coefficient	$m\,s^{-1}$
L	Thickness of particle	m
n	Stirrer speed	revolutions s^{-1}
P	Partition coefficient	—
$[R]$	Reactant concentration	$kg\,mol\,m^{-3}$
Re	Reynolds number	—
Re_i	Reynolds number for stirred system	—
Sh	Sherwood number	—
Sc	Schmidt number	—
t	Time	s
T	Absolute temperature	K
V_{max}	Maximum theoretical rate of enzyme-catalysed reaction	$kg\,mol\,m^{-3}\,s^{-1}$
V_{max}'	Apparent V_{max}	$kg\,mol\,m^{-3}\,s^{-1}$
δ	Boundary layer thickness	m
ψ	Electrical potential	volts
ρ	Density	$kg\,m^{-3}$
ν	Dynamic viscosity	$m^2 s^{-1}$
ϕ	Thiele modulus	—
χ	Porosity	—
τ	Tortuosity	m
μ	Liquid velocity	$m\,s^{-1}$

Chapter 8

Symbol	Interpretation	Units
B	Temperature constant for thermistor	K

Symbols and units

E_G	Standard probe potential	volts
E_G'	Observed probe potential	volts
E_{ref}	Internal reference probe potential	volts
E_{asym}	Asymetric potential	volts
F	Faraday constant	$C\ mol^{-1}$
$[R]$	Reactant concentration	$kg\ mol\ m^{-3}$
R_1	Resistance of thermistor 1	ohms
R_2	Resistance of thermistor 2	ohms
δR	Change in resistance of thermistor	ohms
T	Absolute temperature	K
V	Bridge excitation voltage	volts
v	Bridge output voltage	volts

Chapter 1

Introduction

Historical perspective

It is now becoming common, for authors reviewing aspects of biotechnology, to emphasize the long history of the use of biological systems to bring about desirable chemical conversions. The most widely quoted examples are the conversion of milk to cheese and the fermentation of sugar-containing fluids to alcoholic beverages. It is apparent that our concept of biotechnology has radically altered since these simple processes were first utilized, although the production of bread, cheese and alcohol must still be among the most significant applications. Atkinson (1974) in his text on biological reactors describes the development of biotechnology in terms of three chronological stages:

Pre 1800	some biological processes utilized but ignorance of the mechanisms involved
1800–1900	a period of discovery giving greater understanding of the biological and biochemical basis of bioconversions
Post 1900	a period of industrial development

It is now appropriate to introduce another major era, namely that of genetic engineering.

Post 1970	a period of direct specific biological modification

The historical beginning of enzyme technology stems from the advances made in the period of discovery 1800–1900. During this time, a number of specific chemical conversions were demonstrated using biological tissues, including the decomposition of hydrogen peroxide, degradation of starch to sugars, and the digestion of proteins. The data available in 1836 encouraged Jakob Brezelius to predict that, for unexplained biological reaction mechanisms, 'the future may well reveal them in the catalytic power of the tissues from which the organs of the living body are made up'.

The term 'enzyme' was first introduced in 1878, and early attempts to produce a systematic nomenclature led to the use of the suffix 'ase' appended to the substrate name. During this stage of investigation the concept of specificity, the requirement for coenzymes, and the existence of enzymes in cell-free systems were all established. These findings paved the way for the kinetic description of enzyme activity made by Michaelis and Menten in 1913, and the first preparation of a pure crystallized enzyme, urease, by Sumner in 1926. The significance of these highly purified and highly specific enzyme catalysts first became apparent in the field of chemical analysis, and in the 1930s a number of enzyme-based assays were described.

Although we can use the suffix 'ase' to denote an enzyme name, this system of nomenclature becomes imprecise given the large number of enzymes we now know to exist. In order to standardize nomenclature in a systematic manner a commission was established under the auspices of the International Union of Biochemistry and its reports, with various revisions, are now generally accepted as the basis of enzyme classification. The system is based on dividing enzymes into six main categories depending on the type of reaction catalysed. These categories are given an EC (Enzyme Commission) number and are then further subdivided into categories describing the substrate type and coenzyme requirement. Finally, a number referring to the individual reaction catalysed is assigned. The secondary categories will be defined uniquely according to the primary category to which they are linked.

Given a system of this flexibility, any new enzyme can be classified once its catalytic properties are known. In common usage enzymes are still described by their non-systematic names e.g. L-lactate : NAD^+ oxidoreductase (EC 1.1.1.27) would commonly be called lactate dehydrogenase. In scientific publications it is usual to define fully the enzyme on first reference but for subsequent references to use the International Union of Biochemistry recommended name.

In preparing this book we have used recommended names in the text as far as possible. However, all enzymes referred to are named in full with their EC numbers in Appendix 1.

The use of purified enzymes in larger scale applications was much slower to develop and relied on advances in the techniques of immobilization. Immobilization conferred some stability advantages (described in Chapter 6) but more significantly allowed the expensive pure enzymes to be re-used and retained in a reactor.

Although the possibility of enzyme immobilization was first demonstrated in the early 1900s with the adsorption of yeast extracts onto activated charcoal, it was not until 1953, when techniques for covalently binding enzymes to polyamino–polystyrene resin were developed, that the significance of this technique became

Introduction

apparent. Throughout the 1960s and 1970s there was a vast proliferation of literature describing the use of immobilized enzymes in a range of applications which we will attempt to summarize and explain in the subsequent chapters of the book. It should be stressed, however, that there are alternatives to the use of purified enzymes as biological catalysts. Historically, it has been whole-cell fermentations which have had the most significant impact on the development of biotechnology. Cell immobilization technology has paralleled work on enzymes, and immobilized cells also have a role to play in applied biocatalysis (Klibanov, 1983). More recently, it has been shown that plant tissue and crude tissue homogenates from a range of sources can be used as alternatives to both microbial cells and purified enzymes. Hence for any given application there are a number of alternative biocatalyst preparations from which to choose. It is therefore worth considering what criteria are likely to influence the choice of biocatalyst.

Choice of biocatalyst

Although it is possible to use whole animal or plant tissue to effect bioconversions, it is appropriate in the context of this book to restrict discussion to a comparison of microbial fermentation, immobilized cells and immobilized enzymes. The two most distinctive features of enzymes as catalysts is their high activity and specificity. The advantages that stem from using enzymes for a technological application are that they will occupy little space and can be chosen to react with only one component of a given mixture. This means that the reactor volume will be small and that the production of unwanted byproducts will be minimized. Offset against these advantages is the fact that enzymes are costly to purify and have a limited stability in the purified state. Different enzymes will also have different pH optima, and temperature-related stabilities. If a multistep synthesis requiring several enzymes is envisaged, then these problems will multiply.

Both free and immobilized cell fermentations can be considered as potential competitors for enzyme-based processes. Each method has advantages and disadvantages which affects its suitability for a given application.

Free-cell fermentations are still the most significant group of large-scale biotechnological processes. They are relatively easy to operate and in some cases do not even require a sterile feedstock. As cells are produced as fast as they are washed out of the reactor, there is a constant production of new 'catalyst'. Providing reactor conditions are suitable for growth, the fermentation can be operated at a steady state such that catalytic efficiency does not change. The use of actively growing cells means that fermentations are ideal for the production of complex compounds requiring multistage syntheses. The actively growing cell is able to provide the energy needed for these syntheses from the catabolic breakdown of nutrient compounds. However, the wider range of reactions required for metabolism means an increased likelihood of unwanted byproduct formation. This, coupled with the production of surplus biomass, limits the efficiency of feedstock utilization and hence the economics of the process. The production of surplus biomass and unwanted byproducts will also cause problems in the downstream processing of the product. The desired product must be recovered and purified whereas the unwanted material must be disposed of, all of which adds to the process costs.

Immobilized cells can be regarded as an intermediate stage between fermentation and immobilized enzymes. In some cases cells are killed prior to immobilization and advantage taken of a single enzyme component. Here, a distinction between immobilized cells and enzymes would be a matter of semantics. More generally, immobilized cells would be used where a number of enzyme steps were involved. Energy-consuming syntheses can be performed by immobilized cells but the duration of production will be limited by the chemical energy stored in the cell. Although regeneration is possible, this still represents a major technical difficulty.

Factors affecting enzyme stability will be discussed in greater detail in Chapter 6. However, stability must also be considered in the context of catalyst choice. Ignoring the detailed mechanisms involved in inactivation, it is possible to compare the operational half-lives of immobilized cells and enzymes when used to catalyse the same reaction under similar conditions. It is generally considered that immobilized cells will show a greater stability, which results from the protection afforded the enzymes by the gross cell structure. However, it is clear that this cannot be taken for granted. The reported values for a number of immobilized enzymes and cells, illustrating that the method of immobilization and operating temperature used are critical in determining operational stability, are shown in Table 1.1.

Table 1.1 Reported operational stabilities of immobilized cells and enzymes

Enzyme	Form	Half-life (days)	Temperature (°C)
Aminoacylase	Adsorbed enzyme	40	50
Aspartase	Entrapped *Escherichia coli*	120	37
	Entrapped enzyme	20–25	37
β-Galactosidase (fungal)	Covalently bound enzyme	54	40
Glucoamylase	Covalently bound enzyme	24	55
Glucose isomerase	Heat-treated *Streptomyces*	10–15	70
	Adsorbed enzyme	5–6	70
	Covalently bound enzyme	240	50
	Immobilized enzyme	21–25	65
Histidine ammonia-lyase	Entrapped *Achromobacter*	180	37
Penicillin amidase	Entrapped *E. coli*	17	40
	Covalently bound enzyme	15–25	37

(From Lilly, 1977)

Introduction

Activity and stability of the catalyst are both important in determining the operating costs of the process. The other factor which must be considered is the initial cost of preparation of the catalyst. As will be discussed later, the first choice will be whether or not to immobilize the catalyst. In making this decision the cost of immobilization will be an important factor, coupled with the resultant operational costs of the process (Fig. 1.1). A comparison of their merits shows that immobilized catalysts offer a considerable saving in both labour and materials compared with batch systems. If an immobilized catalyst is chosen, then the final comparison must be made between cells and enzymes based on the relative costs of preparation (Lilly, 1977). In a simplistic assessment we could assume that the support costs and associated immobilization procedures would be approximately equal, which would leave the additional cost of extraction and purification of the desired enzyme to be accounted for.

The major restriction on the use of enzymes stems from the technical difficulties arising from arranging multi-enzyme systems, and the difficulty in providing the energy required to drive thermodynamically unfavourable synthetic reactions. Approaches to solving both problems have been demonstrated in the laboratory but, apart from isolated pilot-scale studies, these have yet to be scaled up. In the cell, the driving energy for synthetic reactions is provided by coenzymes which act as chemical mediators and, in the normal course of events, these would be regenerated. One of the most important of these coenzyme mediators from a technological point of view is nicotinamide adenine dinucleotide which can exist in an oxidized or reduced state ($NAD^+/NADH$). This coenzyme provides reducing or oxidizing potential in a wide range of reactions. Although large-scale processes involving NAD regeneration have yet to be developed, it has been found that the enzyme can be replaced by a redox dye or an electrode surface. This allows the formation of a bioelectronic sensor, where electrical activity is a function of substrate concentration.

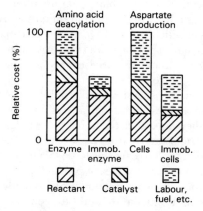

Fig. 1.1 Relative operating costs of two processes showing the effects of catalyst immobilization on the breakdown of costs.
(From Lilly, 1977)

It is fair to say that, apart from small-scale applications like enzyme sensors, most current commercial processes using enzymes are restricted to reactions where there is no coenzyme requirement. However, this should be seen as a technical rather than a fundamental problem. Further developments may well lead to processes using multi-enzyme synthesis with coenzyme regeneration *in situ*. The most significant factor influencing the choice of process and catalyst will be process economics. As the technology improves, some of the disadvantages mentioned will be overcome and the methods of choice for a given application will change (Katchalski-Katzir and Freeman, 1982).

Legal implications in the use of enzymes

The criteria of economic viability for a process is intrinsically linked with the legislative guidelines that apply to its operation and its product purity (Reichelt, 1983). Most legal restrictions apply to matters of safety, both in terms of plant operation and product specification. The cost of satisfying the appropriate legal requirements must obviously be considered when process costs are being determined. In the case of biological materials there are several obvious areas of potential hazard:

(1) Microbiological
(2) Chemical toxicity
(3) Activity-related toxicity
(4) Allergenicity

Problems concerning microbiological activity will relate to the source organisms chosen as a raw material, and will be reflected in the safety measures incorporated into the production process. Products aimed at the consumer market will be required to meet purity criteria which demonstrate that they are not subject to harmful chemical or microbial contamination.

Chemical toxicity usually arises from contamination by microbial secondary metabolites, examples of which include mycotoxins and aflatoxins. They can arise in a wide range of food materials as a result of microbial contamination. A greater awareness of the long-term consequential effects of these toxins is leading to a tightening of legislative controls.

Activity-related toxicity is best illustrated by the example of proteolytic enzymes used in the production of biological washing powders. If these enzymes are inhaled as a dust, they can attack the lining of the lungs, causing irritation and respiratory disorders. Highly active proteinases can have degrading power similar to strong caustic solutions and so care has to be taken, particularly with powders prior to their dilution in the final stages of production.

In the early stages of enzyme production proteins tended to be handled as fine dusting powders, leading to inhalation and skin contact by process workers. In addition to activity-related problems, this also led to allergic reactions in a proportion of the workforce. As with hay fever, the body reacts to the presence of a

Introduction

foreign biological material by mounting an immune response. The response will tend to become more severe with subsequent re-exposure and can be dangerous or even fatal in extreme cases.

Having identified potential hazards associated with enzyme usage and production, it is useful to look at the general approach which has been taken to control their use. As legislation is continuously under review, it is not appropriate to give more than a general outline of the philosophy in a text of this kind, and only the UK position has been considered, although other countries have similar policies.

In 1982, Her Majesty's Stationery Office (HMSO) published a report, *Review of Enzyme Preparation*, by the Food Additives and Contaminants Committee of the Ministry of Agriculture, Fisheries and Food. In this report enzymes were classified into five groups depending on their suitability for use in the food industry taking into account the differing types of toxicity as outlined above.

Group A Substances that the available evidence suggests are acceptable for use in food

Group B Substances that on the available evidence may be regarded as provisionally acceptable for use in food, but about which further information must be made available within a specified time for review

Group C Substances for which the available evidence suggests toxicity and which ought not to be permitted for use in food until adequate evidence of their safety has been provided to establish their acceptability

Group D Substances for which the available information indicates definite or probable toxicity and which ought not to be permitted for use in food

Group E Substances for which inadequate or no toxicological data are available and which it is not possible to express an opinion as to their acceptability for use in food

In an attempt to comply with the regulations, some industrial associations have drawn up their own guidelines. In the case of food enzymes, the Association of Manufacturers of Animal and Plant Food Enzymes have developed a pragmatic approach based on the degree of prior usage of the source material in the food industry (Table 1.2). The organisms mostly used for enzyme production are categorized as outlined in Table 1.3. The problems of toxicity in the pharmaceutical industry are naturally much greater, and different rules apply. For products not designed to be ingested or injected into the body the toxicological criteria will be much less severe. However, the problems associated with safe handling will be common to all enzyme manufacture.

An increasing awareness of the potential for enzyme powders to cause allergic responses in process workers has led to changes in the formulation of products. Most manufacturers have switched to liquid formulations to avoid the problems of dust. Where a powdered product is required, techniques such as encapsulation

Table 1.2 Safety testing of food enzymes based on the AMFEP classification

Group Tests (X = to be performed)	A, Micro-organisms that have traditionally been used in food, or in food processing	B, Micro-organisms that are accepted as harmless contaminants present in food	C, Micro-organisms that are not included in A or B
Pathogenicity	In general no testing required		X
Acute oral toxicity, mouse and rat		X	X
Subacute oral toxicity, rat four weeks		X	X
Three month oral toxicity, rat		X	X
In vitro mutagenicity		X	X
Teratogenicity, rat			(X)*
In vivo mutagenicity, mouse and hamster			(X)*
Toxicity studies on the final food			(X)*
Carcinogenicity, rat			(X)*
Fertility and reproduction			(X)*

* Only to be performed under exceptional conditions.

(From Reichelt, 1983)

Table 1.3 Micro-organisms used for enzyme production

Group A	Micro-organisms that have traditionally been used in food or in food processing
	Bacillus subtilis (including strains known under the names *mesentericus, natto* and *amyloliquefaciens*)
	Aspergillus niger (including strains known under the names *awamori, foetidus, phoenicis, saitoi* and *usamii*)
	Aspergillus oryzae (including strains known under the names *sojae* and *effesus*)
	Mucor javanicus
	Rhizopus arrhizus
	Rhizopus oligosporus
	Rhizopus oryzae
	Saccharomyces cerevisiae
	Kluyveromyces fragilis
	Kluyveromyces lactis
	Leuconostoc oenos
Group B	Micro-organisms that are accepted as harmless contaminants present in food
	Bacillus stearothermophilus
	Bacillus licheniformis
	Bacillus coagulans
	Bacillus megaterium
	Bacillus circulans
	Klebsiella aerogenes
Group C	Micro-organisms that are not included in groups A and B
	Mucor miehei
	Mucor pusillus
	Endothia parasitica
	Actinoplanes missouriensis
	Streptomyces albus
	Bacillus cereus
	Trichoderma reesei (T. viride)
	Penicillium lilacinum
	Pencillium emersonii
	Sporotrichum dimorphosporum
	Streptomyces olivaceus
	Penicillium simplicissium
	Penicillium funiculosum

and granulation are used to minimize dust formation. The expansion in the use of immobilized enzymes has also tended to minimize these problems.

The legislative aspects of production are covered in the UK by the Health and Safety at Work Act 1974, which is designed to ensure the health, safety and welfare of employees. As a result of the introduction of the legislation, good manufacturing practice is widely adopted and it has been suggested that quality control criteria applying to both raw materials and products in the food enzyme industry are now generally as high as those in the pharmaceutical industry.

Growth of the enzyme industry

To some extent the expansion of the enzyme industry mirrors the development of our scientific understanding of enzymes. The advent of enzyme technology can be traced back to Christian Hansen who marketed the first standardized enzyme preparation (rennet) for a technological application (cheese manufacture) in 1874. Subsequent developments included the awarding of a US patent covering the use of diastatic enzymes in 1894, and a patent for the use of pancreatic proteinases for bating hides in 1907. These were followed by patents for the chill-proofing of beer using proteolytic enzymes (1911) and bacterial enzymes for use in desizing textiles (1917). The situation up to 1977 is summarized in Table 1.4,

Table 1.4 Commercially important enzyme preparations

Source	Name	Commercially available before 1900	1950	1976	Current production (tons of enzyme protein per year)
Animal	Rennet	X			2
	Trypsin		X		15
	Pepsin		X		5
Plant	Malt amylase	X			10 000
	Papain		X		100
Microbial	Koji	X			?
	Bacillus proteinase		X		500
	Amyloglucosidase			X	300
	Bacillus amylase		X		300
	Glucose isomerase			X	50
	Microbial rennet			X	10
	Fungal amylase	X			10
	Pectinase		X		10
	Fungal proteinase	X			< 10

(From Aunstrup, 1977)

Introduction

although it must be realized that production figures may not be directly comparable due to purity and activity variations between different enzyme preparations.

By the early 1970s, enzyme technology was entering into a period of industrial process development, the production of amino acids, and of sweeteners based on isomerized glucose. At this time both the European and American markets were dominated by proteinase enzymes used in the detergent industries (Fig. 1.2), although there was speculation that the market for enzymes used in the food industry would show a major increase (Dunnill, 1980; Lewis and Kristiansen, 1985).

An examination of the figures reported by Aunstrup (1977) for the world enzyme market in 1982 (Table 1.5) show that this assumption was correct. A comparison of the annual production figures for glucose isomerase shows an increase from 50 to 1500 tons between 1977 and 1982, clearly demonstrating why the EEC imposed tariffs on high fructose corn syrup to protect the sugar market.

In 1981 the world market for enzymes was reported at $200 million; in 1985 the US Office of Technology assessment suggested a world market of $250 million. As the figures are certain to vary with source, the safest interpretation is that the market for industrial enzymes showed a dramatic increase throughout the 1970s with the development of a number of high volume users in the food industry. This expansion has now slowed and current increases are more likely to be attributed to

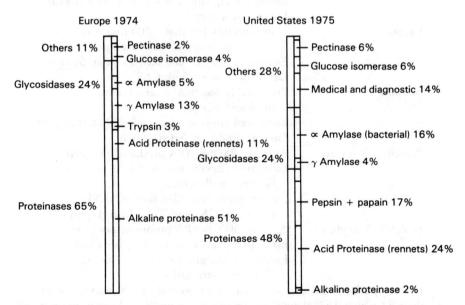

Fig. 1.2 Breakdown of enzyme sales for Europe and the United States during the mid 1970s.
(From Solomons, 1977)

Table 1.5 Immobilized enzymes—annual product consumption 1982 worldwide

Immobilised enzyme product (listed alphabetically)	Size
Aminoacylase	Presumably less than 250t L-amino acid produced per year
	Estimated enzyme amount: less than 5t/year
	Largest producer: Amano, Japan (?)
Amyloglucosidase/ glucoamylase	Presumably less than 5000t syrup produced per year
	Estimated enzyme amount: less than 1t/year
	Largest producer: Tate & Lyle, UK
Glucose isomerase	Approx. 2150000t 42% HFCS and 1450000t 55% HFCS produced per year (after Poulsen 1981–82)
	Estimated enzyme amount: 1500–1750t/year
	Largest producers: Novo Industri, DK and Gist Brocades, NL
Hydantoinase	Presumably less than 50t p-phenylglycine produced per year
	Estimated enzyme amount: less than 1t/year
	Largest producer:?
Lactase	Presumably less than 1000 tons d.s. lactose hydrolysates produced per year
	Estimated enzyme amount: less than 5t/year
	Largest producer: Valio, SF
Nitrilase	Presumably less than 5t acrylamide produced per year
	Estimated enzyme amount: less than 0.1t/year
	Largest producer: Nitto, Japan
Penicillin G acylase	Approx. 4000t 6-APA produced per year
	Estimated enzyme amount: 3–4t/year (after Godfrey & Reichelt)
	Largest producers: Gist Brocades, NL, Beecham, UK, and Toyo Jozo, Japan
Penicillin V acylase	Approx. 500t 6-APA produced per year
	Estimated enzyme amount: approx 1t/year
	Largest producers: Biochemie, Austria and Novo Industri, DK

For comparison, the largest enzyme for once-only use is amyloglucosidase, of which 15,000 tons are produced annually (substrate approx. 20×10^6t dry matter) and used for syrup, ethanol, and beer production.

(From Poulsen, 1984)

high-value, low-volume materials for use in the medical and related fields. In the future the market may be expected to show further growth when enzymes start to be introduced into processes for industrial chemical production (Bjurstrom, 1985; Harnisch and Wöhner, 1985).

Chapter 2

Commercial sources of enzymes

Introduction

The current edition of the *International Union of Biochemistry Handbook of Enzyme Nomenclature* (Webb, 1984) lists nearly 2500 different enzyme-catalysed reactions which have been reported in the scientific literature. This figure is an underestimate of the total number of enzymes discovered as there is often a multiplicity of proteins, each with slightly different properties, that can catalyse each of the reactions. Of these 2500 enzyme 'types', some 15% are available from biochemical suppliers in quantities ranging from micrograms to kilograms and they are provided essentially for research purposes. However, only about forty to fifty enzymes are produced on the industrial scale, i.e. multikilograms to tons per annum. These forty or so enzymes are produced from microbial, plant and animal sources and in general they catalyse simple hydrolytic reactions (Godfrey and Reichelt, 1983). The ability to utilize coenzyme-requiring biosynthetic enzymes remains a major challenge to the industry but is one that could result in large financial rewards if solved.

Sources of enzymes

Most of the commercially important enzymes have been produced from a limited range of micro-organisms. The organisms used have been dictated largely by precedent and have been restricted largely to those that have been used in the food industry or are found as normal harmless contaminants in foodstuffs (see Chapter 1). This apparently restrictive approach has been forced upon the industry by the

Commercial sources of enzymes

immense cost of obtaining approval for the use of novel organisms by the appropriate regulatory authorities.

Traditionally a few enzymes have been obtained from animal sources, e.g. pancreatic lipase and trypsin, although these may be replaced by similar enzymes derived from micro-organisms. However, the microbial replacements, whilst catalytically efficient, show subtle differences in properties that can be crucial to the process application. This is particularly well illustrated with the use of proteolytic enzymes in cheese making (see Chapter 7). Consequently, recent work in this area has led to the use of recombinant DNA techniques (see later in this chapter) to enable mammalian genes to be cloned into suitable bacteria or yeasts to facilitate production of animal enzymes using conventional fermentation technology.

Plants have also been a traditional source of a limited number of enzymes. In particular, the cysteine proteinases have been isolated from the latex produced by papaya, fig, pineapple and, to a lesser extent, other plants. These plants have the advantage that it is easy to obtain the latex, principally from the fruit, using unskilled labour. The latex is usually allowed to dry in the sun and can if necessary be used directly as a crude enzyme preparation. The latex from these plants often contains very large quantities of enzyme. For example, approximately 90% of the protein in the gum derived from *Ficus glabrata* latex comprises a mixture of ten very similar proteolytic enzymes. This must rate as one of the richest sources of a particular type of enzyme activity. Similarly, the fresh latex from *Carica papaya* fruit has a number of proteolytic enzyme activities of which one of these, papain, constitutes 7% of the total soluble matter. Clearly, these are excellent sources of enzymes and, furthermore, they can be obtained without damaging the plants or the very edible fruit.

Another major source of plant-derived enzymes is malted barley. Traditionally, the brewing industry has used this material as a crude source of various proteinases and glycohydrolases. In terms of bulk usage malted barley makes an important contribution to the enzyme market.

The use of enzymes from both plant and animal sources represents a declining proportion of the total enzyme market although the reasons are different in each case. The plants that produce proteolytic enzymes grow best in tropical or subtropical regions and are therefore produced mainly by Third World countries. Although there is the advantage that labour is cheap, this is often counterbalanced by a poorly organized infrastructure and inherent political and economic instability. In contrast, enzymes derived from animal sources are governed by other economic problems. Essentially, the production of enzymes, usually from offal, is closely tied to the meat industry. Consequently, the enzyme manufacturers are often in the situation of accepting raw materials at prices fixed by factors external to the industry. This, coupled to the inherent problems of using animals for purposes other than food, places major limitations on animals as a source of enzymes.

Consideration has been given to the production of enzymes from the culture of either animal or plant cells. Technically, the growth of animal or plant cells in large quantities is now established. However, economically it seems unlikely that this will be a significant source of enzymes in the immediate future. Basically, the growth of these cells in culture is slow and therefore it is diffcult to maintain sterility. Currently, it is estimated that, to be viable economically, an enzyme from cell culture must have an inherent value one thousand times greater than an equivalent product produced by traditional microbial fermentation.

The main result of the problems in using either plant- or animal-derived products is the establishment of micro-organisms as the major source of industrial enzymes.

Microbial enzymes

Micro-organisms produce a tremendous range of potentially useful enzymes and furthermore many of these are secreted from the cell. Also, micro-organisms are quick and easy to grow in culture and the technology of scale-up is well established.

The advantages of extracellular enzymes are three-fold. First, as the enzyme is secreted from the cell this obviates the need for cell-breakage techniques which are often difficult to employ on a large scale. Second, only a limited number of proteins are secreted and so it is relatively easy to isolate the desired enzyme from the mixture. In comparison intracellular enzymes must be separated from a whole range of other proteins and contaminating materials, e.g. nucleic acids. Third, extracellular enzymes tend to have a more robust structure and are less susceptible to denaturation than their intracellular counterparts. Therefore, micro-organisms, particularly those that secrete enzymes, are the preferred 'raw-material' of the industrial enzymologist (Holland et al., 1986).

Conventionally, extracellular enzymes are defined as those that have crossed the cell membrane. This point is important because micro-organisms possess a variety of cell-wall structures and in many cases the extracellular enzyme is associated with these rather than released into the culture medium.

During synthesis within the bacterial cell certain proteins are recognized as being intended for export by the presence of an N-terminal signal peptide. These extracellular enzymes are partially synthesized on the free ribosomes in the cytosol before attachment of the complex (directed by the N-terminal signal peptide) to the cell membrane (Davis and Tai, 1980). Biosynthesis of the enzyme then continues with concomitant secretion of the growing polypeptide chain (co-translational secretion). During the process of secretion the N-terminal peptide is removed by a membrane-bound proteinase (Fig. 2.1).

In fungi, and other eukaryotes, the process of secretion is essentially the same as that described for bacteria. However, during synthesis the peptide chain is pushed through into the lumen of the endoplasmic reticulum rather than directly through the cytoplasmic membrane (Walter et al., 1984). From the lumen of the endoplasmic reticulum the enzyme is processed through the Golgi apparatus and it is finally exported to the cell membrane in vesicles. This extra layer of complexity in eukaryote cells allows the secreted enzymes to be modified extensively during passage through the endoplasmic reticulum and Golgi apparatus (Fig. 2.2). These post-translational modifications (Freedman and

Commercial sources of enzymes

(a) *Formation of ribosome/mRNA initiation complex*

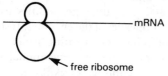

(b) *Start of protein biosynthesis*

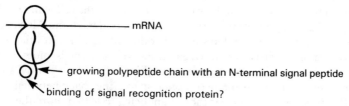

(c) *Signal peptide directed attachment of ribosome to plasma membrane*

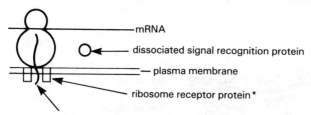

peptide chain continues to be synthesized and is pushed through the membrane

(d) *Removal of signal peptide*

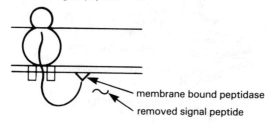

(e) *End of protein biosynthesis*

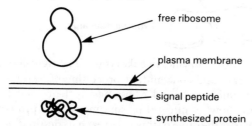

* Although signal recognition and ribosomal receptor proteins have been identified in eukaryotes their presence in prokaryotes has only been inferred from genetic studies.

Fig. 2.1 Co-translational secretion of bacterial proteins.

Hawkins, 1980), which may include glycosylation, removal of small peptide fragments, introduction of disulphide bridges and many other possibilities, usually have important influences on the stability and properties of the enzymes. It should be mentioned that although this is the general mechanism operating in eukaryotes there are instances of certain fungi that do not possess a conventional Golgi apparatus. The mechanism of secretion in those organisms is unknown.

Bacteria and fungi are surrounded by cell walls that are composed of peptidoglycan and $(1 \rightarrow 3)$-linked glucans/chitin complexes respectively. Once the enzyme has been secreted it is usually free to diffuse across the cell wall and, in the case of fungi and Gram positive bacteria, into the external medium. Alternatively, some enzymes will remain associated with either the cell membrane or the cell wall structure. Secretion of enzymes in Gram negative bacteria is more complex because of the presence of an outer membrane (Fig. 2.3). Consequently, secreted enzymes are often retained in the periplasmic space (the space between the cell wall and the inner membrane) rather than being released into the culture medium. This extra membrane and the different localization of proteins has

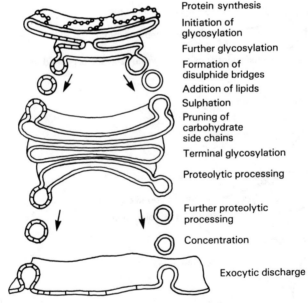

Fig. 2.2 The secretion and processing of enzymes in eukaryotes. The right-hand side of the diagram shows the itinerary and concomitant processing of secretory proteins from their assembly by membrane-bound ribosomes to their exocytic discharge. The left-hand side shows how an integral plasma-membrane protein made in the endoplasmic reticulum uses the same pathway to reach its destination. (From de Duve, 1985)

Commercial sources of enzymes

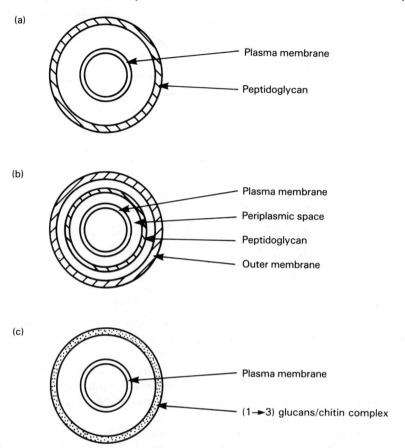

Fig. 2.3 Diagrammatic representation of typical cell wall structures: (a) Gram positive bacterium; (b) a Gram negative bacterium; (c) a fungus.

restricted the use of Gram negative organisms for the production of extracellular enzymes. In fact, the only major enzyme produced from a Gram negative bacterium is pullulanase from *Klebsiella pneumoniae*.

Control of microbial enzyme production

The optimization of enzyme production from micro-organisms is dependent on a number of interrelated factors. In practical terms this means that an enzyme may be synthesized only during part of the growth cycle. For example, some proteins are synthesized and secreted during the active growth phase (exponential growth) of the organism whereas others do not appear until the stationary phase (Fig. 2.4).

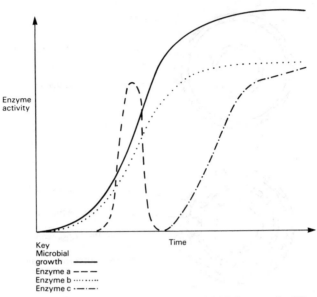

Fig. 2.4 Enzyme synthesis as a function of microbial growth. Three typical examples of enzyme production by micro-organisms are illustrated, although it should be emphasized, that any intermediate situation is possible.

An added complication is that a particular enzyme may only be synthesized when the bacteria are growing in a medium of appropriate composition. For example, the presence of simple carbon compounds such as D-glucose may repress synthesis of an enzyme.

To understand why there is a variation in the onset and extent of synthesis it is necessary to know a little about the genetic control of enzyme production (Booth and Higgins, 1986). Essentially, there are two identifiable classes of enzymes. The first group comprises the constitutive enzymes. Numerically, these form a minor proportion of the total but they are usually enzymes that are involved in central aspects of metabolism. A constitutive enzyme is produced continuously regardless of whether or not substrate is present. The second, much larger, group comprises the inducible enzymes. These enzymes are produced at very low but significant levels until a suitable substrate becomes available to the organism. The substrate itself, or degradation products derived from it, may then induce a greatly increased synthesis of the enzyme. It is quite common for synthesis of an induced enzyme to increase by a thousand-fold or more under these circumstances. Induction can be a difficult phenomenon to follow experimentally, as the inducer is often metabolized by the induced enzyme. It is best to study induction by the use of gratuitous inducers. These are compounds that induce certain enzymes but are not metabolized. For example, the disaccharide sophorose ($\beta 1 \rightarrow 2$ glucobiose) induces the production of cellulose-degrading enzymes in certain fungi yet it is not metabolized itself.

Regardless of whether an enzyme is constitutive or inducible, there is a further control mechanism that can operate. In the presence of a readily assimilated carbon source, e.g. D-glucose, the synthesis of many enzymes is repressed. The rationale behind this phenomenon is that an organism can utilize glucose without the need to expend energy synthesizing enzymes for the degradation of more complex carbon sources. Carbon-catabolite repression is a commonly occurring feature of micro-organisms and many constitutive and inducible enzymes are affected in this way. This can make it very difficult to decide whether the transient synthesis of an enzyme is a result of induction, repression or a combination of these mechanisms. The synthesis of other enzymes may be repressed by the presence of nitrogen-containing compounds, sulphur-containing compounds, and so on. In general, a particular form of repression is most likely to affect those enzymes involved in the metabolism of compounds containing the relevant element. For example, the synthesis of the extracellular proteinase of *Bacillus subtilis* is repressed by the presence in the culture medium of certain nitrogen-containing compounds, e.g. glutamate or aspartate.

Although it is not appropriate to describe the detailed mechanisms of induction and repression, it is perhaps worth considering a simple model system. One of the simplest and best understood mechanisms is the induction and repression in *Escherichia coli* of the enzymes involved in lactose metabolism. The genes, *lac* z, *lac* y and *lac* a for the three proteins β-galactosidase, galactoside permease and thiogalactoside transacetylase are clustered together on the bacterial chromosome to form the *lac* operon. The transcription of these three genes is regulated via the control region of the operon which consists of the *lac* i gene and the promoter and operator regions (Fig. 2.5).

It is the efficiency with which the enzyme DNA-dependent RNA polymerase is able to bind to the promoter that governs the efficiency of transcription. The *lac* i

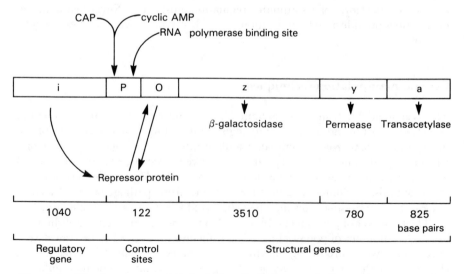

Fig. 2.5 The *lac* operon of *Escherichia coli*. CAP = catabolite activator protein.

gene encodes the *lac*-repressor protein which is synthesized constitutively at very low levels. This tetrameric protein will bind tightly to the operator region and prevent RNA polymerase from transcribing the genes, *lac* z, y and a. Lactose and other β-galactosides act as inducers of enzyme synthesis by binding to each of the four subunits of the *lac*-repressor protein, causing a conformational change, and preventing binding to the operator (the gratuitous inducer isopropylthiogalactoside is commonly used in these experiments). The RNA polymerase is now able to transcribe the genes that encode the structural proteins. In other words, the enzymes of lactose metabolism have been induced (strictly speaking, derepressed) by the presence of β-galactosides.

The *lac*-operon is also subject to carbon-catabolite repression. In the presence of D-glucose the *lac* operon is switched off regardless of whether lactose is present in the growth medium. This repression mechanism operates through cyclic AMP (cAMP) and the catabolite activator protein (CAP). The binding of RNA polymerase to the operon is enhanced by the attachment of a cAMP·CAP complex to part of the promoter. In the presence of D-glucose the level of cAMP in the cell is greatly reduced and the CAP protein dissociates from the DNA. This results in a much reduced binding of RNA polymerase to the promoter with a consequent reduction in the transcription of the three structural genes. It is not clear how the presence of D-glucose reduces the level of cAMP in the cell although it is known to be an indirect effect dependent on a metabolite of the sugar.

Although the *lac* operon provides a useful model of how enzyme synthesis may be controlled, it is not a mechanism that is generally applicable to all microorganisms. For example, in many bacteria it has been impossible to detect the presence of cAMP, so clearly a different mechanism must operate in these circumstances. Also, the genes encoding enzymes involved in a single pathway may be situated far apart from each other in the bacterial chromosome. In those cases some other form of co-ordinated regulation must occur. Nevertheless, the *lac* operon does provide a useful indication of how bacterial enzyme synthesis may be controlled.

Genetic manipulation techniques

The recent advances in recombinant DNA technology offer many potential advantages to the industrial enzymologist. For example, it is now possible to incorporate a gene from an animal or a plant into a micro-organism. Therefore, enzymes that may be difficult to produce by conventional means may be harvested with ease by applying traditional fermentation technology to the recombinant micro-organisms. Similarly, genes for enzymes from pathogenic or otherwise undesirable micro-organisms may be transferred to a more acceptable host.

The basic techniques of recombinant DNA technology are well established but there are a number of potential practical difficulties that need to be considered, particularly when producing enzymes from eukaryote genes (see later in this chapter). Another point that should be emphasized is that genetic manipulation experiments are subject to strict legal regulations in most countries and that a

Commercial sources of enzymes

particular cloning strategy must be approved by an appropriate committee before work commences. Even fairly straightforward cloning experiments that may be acceptable on the laboratory scale will require more stringent containment on scale-up.

Gene cloning requires the use of a number of relatively simple techniques (Old and Primrose, 1985). These are the ability to (i) cut and join DNA molecules, (ii) introduce the recombinant DNA into a host organism using suitable cloning vectors, e.g. plasmids, cosmids or λ phage derivatives, and (iii) identify the desired clones. The particular strategy that is employed depends on a variety of factors and some of these are considered below.

One of the simplest strategies is the transfer of a gene from one strain of bacterium to another by shot-gun cloning (Fig. 2.6). The DNA from the donor organism is isolated and then partially digested with a Type II restriction endonuclease of appropriate specificity. Restriction endonucleases recognize specific sequences of nucleotides (commonly hexanucleotide sequences) within DNA molecules and cut both strands in either a staggered or a blunt-ended fashion (Fig. 2.7). If the enzyme makes a staggered cut, the resultant pieces of DNA have short single-stranded 'sticky ends' that are capable of base-pairing with other fragments. A suitable cloning vector such as a plasmid is cut with the same enzyme. Combination, in the appropriate stoichiometric quantities, of the cut vector and the many fragments of the donor DNA will allow random reassociation of molecules to occur by base pairing of the 'sticky ends'. Among the many species of recombinant molecules that will be formed should be some in which the vector is combined with the gene of interest. The missing phosphodiester linkages in the DNA backbone may be reformed with the aid of DNA ligase, a process known as ligation.

The next stage is to persuade the host bacteria to incorporate the recombinant molecules. The process that is used is dependent on the type of vector and the simplest type to use is a plasmid. Plasmids are double-stranded circular pieces of DNA, capable of replication, that are present in addition to the chromosomal DNA in certain bacteria. One class of plasmids encodes single or multiple antibiotic resistance and it is some of these that form the basis of cloning vectors. Bacteria may be persuaded to take up plasmids by a process known as transformation. Treatment of bacteria with cold calcium chloride followed by a short heat shock at 42 °C will cause a small proportion of the exogenously added plasmid or recombinant plasmid to be assimilated by the host cells. This is a fairly inefficient process and typically less than 0.1% of the bacterial population can be persuaded to accept and retain the plasmid DNA molecules. Also, smaller DNA molecules are incorporated preferentially into the host cells. This can be a problem as recircularized plasmids which do not contain insert DNA will be transformed in preference to the recombinant molecules. However, these problems can be overcome by the use of suitable selection procedures.

It is important to be able to screen rapidly for the desired clone. A simple calculation shows that the proportion of cells containing the desired gene will be exceedingly small. For example, if a typical bacterial chromosome that contains the target gene is 5×10^6 base pairs long and the restriction enzyme recognizes a

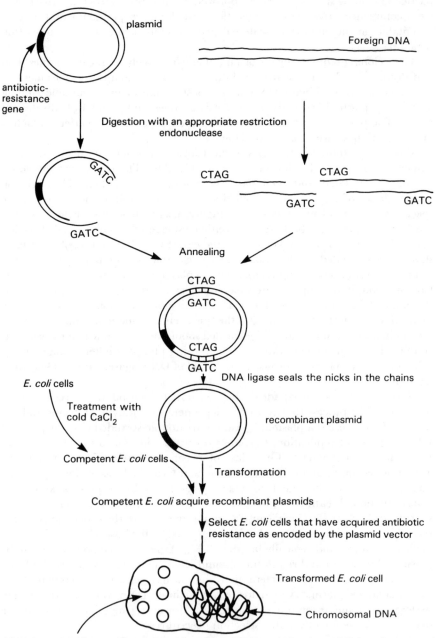

Fig. 2.6 Shot-gun cloning of DNA into a plasmid vector.

Commercial sources of enzymes

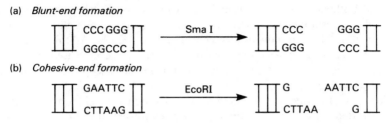

Fig. 2.7 Mode of action of some restriction endonucleases. The enzyme Sma I is derived from the organism *Serratia marcescens*, and EcoRI from *Escherichia coli*. Cohesive ends are also known as 'sticky ends'.

specific hexanucleotide sequence, i.e. it will cut DNA on average 1 in 4^6 (1 in 4096) bases, approximately 1200 fragments will be generated. It is likely that only one of these 1200 fragments will contain the required gene. This fact combined with the observed efficiency of transformation of 0.1% means that less than one bacterium in 1.2×10^6 will contain the appropriate recombinant molecule. Furthermore, each DNA fragment may be inserted into the vector in either of two directions. Usually insertion in only one of these two directions will result in gene expression. Therefore, the number of bacteria containing a correctly expressing recombinant plasmid will be 2.4×10^6.

Selection of the appropriate clone is achieved in two stages. Most plasmids that are used as cloning vectors contain two or more antibiotic-resistance genes. For example, the commonly used plasmid pBR 322 contains ampicillin- and tetracycline-resistance genes (Fig. 2.8) and will confer resistance against both antibiotics to the host bacteria. If foreign DNA is inserted into the Pst 1 restriction site of pBR 322, then the recombinant molecule will only confer resistance to tetracycline as the ampicillin gene has been inactivated. This is the technique of insertional inactivation. Incubation of the transformed organisms on agar plates containing tetracycline will ensure that only those bacteria that have aquired

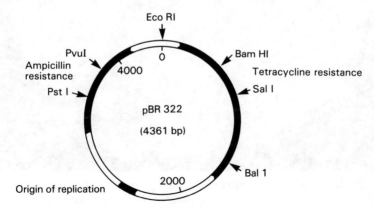

Fig. 2.8 Map of plasmid pBR 322. The antibiotic-resistance genes are shown, together with a selection of unique restriction enzyme sites.

either pBR 322 or a recombinant molecule will survive. Further selection of those tetracycline-resistant clones which are not able to grow on ampicillin will select for those bacteria that contain only recombinant molecules. At this stage in the process the chances of selecting the correct clone have increased from one in 2.4×10^6 to one in 2.4×10^3. This number is much easier to cope with at the second stage of screening.

The type of second stage screening that will be used to identify a particular clone will depend on whether the enzyme is expressed in the new organism. If expression does occur, then screening is simply a matter of devising a plate assay. For example, bacteria producing polymer-degrading enzymes will produce clearing zones when grown on agar plates containing an appropriate substrate (Fig. 2.9). Transformants containing a recombinant plasmid in which the enzyme is not expressed or readily measured can be identified by hybridization. Bacterial clones are patched out from a master plate onto nitrocellulose filter overlays on

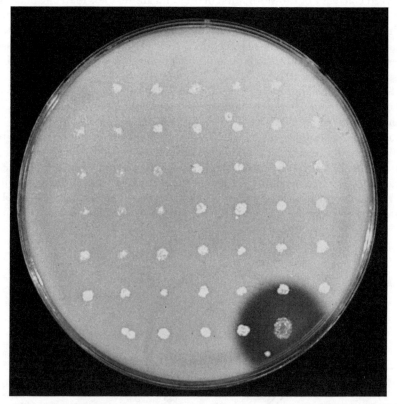

Fig. 2.9 Identification of enzyme-secreting bacteria. Colonies of *Klebsiella pneumoniae* have been plated out onto alginate-containing solid medium. Alginate lyase producing organisms produce a clearing zone around the colony, which may be made visible by staining with cetylpyridinium chloride.

Commercial sources of enzymes

agar plates. After growth the cells are lysed and the released DNA binds strongly to the filter. A radioactively labelled DNA or RNA probe that is complementary in sequence to the appropriate gene is allowed to base pair. Those colonies that contain the gene, and are therefore radioactively labelled, may be detected by autoradiography.

An alternative to cloning in plasmids is to use either λ phage derivatives or cosmids as vectors. The bacteriophage λ infects *E. coli* and in the process inserts viral DNA into the bacterial cell with considerable efficiency. Clearly, if a means can be found of inserting foreign DNA into the viral genome, then the recombinant DNA will be efficiently transferred to *E. coli*. Derivatives of λ phage, eg. Charon 4A, are available in which much of the central portion of the genome has been deleted without compromising the ability to act as a cloning vector. This 'missing' portion may be replaced with foreign DNA using the ligation techniques described previously. Providing that the overall length of the recombinant DNA is between 78% and 105% of the normal λ genome (49 kb), then it can be packaged *in vitro* into empty phage heads (Fig. 2.10). A culture of *E. coli* may then be infected with the phage particles and the recombinant DNA will be transferred efficiently into the bacteria by the process of transfection.

There is a two-fold advantage in using phage vectors rather than plasmids. First, with derivatives of λ phage such as Charon 4A, large insert sizes of 8–22 kb can be incorporated, thus reducing the total number of clones that need to be screened initially. Second, if the two 'arms' of the viral genome combine without foreign DNA, the overall length of the molecule will be too small to package. Therefore, only recombinant DNA molecules should be cloned. Once the gene for an enzyme has been successfully cloned it is necessary to sub-clone into plasmids to reduce the size of the insert DNA. This helps to improve the stability of the cloned DNA within the cells and ensures that spurious proteins coded by other parts of the recombinant molecule are not expressed.

Cosmid vectors are essentially plasmids that contain the *cos* site. The *cos* site is a short region of DNA (12 bp) derived from the λ phage genome that allows the cosmid to be packaged into phage heads *in vitro*. However, once the cosmid has been transfected into bacteria, it behaves as a plasmid and is thus easier to deal with in practical terms, especially from the point of view of sub-cloning. Typically, cosmids may contain insert DNA within a size range of 20–40 kb. Cosmids have the advantage of both plasmid and λ-phage vectors.

The cloning of a gene from a bacterium is usually a straightforward operation. However, problems may arise when eukaryote genes are cloned into bacteria. Eukaryote genes usually contain non-coding sequences (introns) which are spliced out after transcription of the DNA. Unfortunately, bacteria (with the exception of the Archaeobacter) are unable to remove introns and consequently direct transfer of an eukaryote gene into a prokaryote will seldom result in successful expression of the protein. This process can be overcome by the process of reverse transcription (Fig. 2.11). In this case the starting point is eukaryote mRNA; this has already been spliced appropriately and the coding sequence is uninterrupted. The mRNA then acts as a template for the enzyme RNA-directed DNA polymerase (reverse transcriptase) which synthesizes a complementary strand of DNA. The

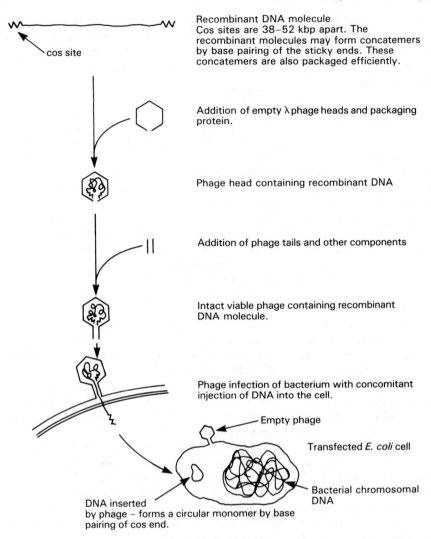

Fig. 2.10 Cloning of DNA using λ phage vectors.

mRNA : DNA hybrid molecule is now treated with RNAase H and DNA-directed DNA polymerase I to enable the mRNA chain to be replaced by a strand of DNA. The RNAase H produces random nicks in the phosphodiester linkages of RNA and then the DNA polymerase I removes the ribonucleotides ($5' \rightarrow 3'$ exonuclease function of DNA polymerase I) and replaces them with the corresponding deoxyribonucleotides in a process known as 'nick translation'. The complementary DNA (cDNA) molecule contains a continuous coding sequence for the protein and is therefore different from the genomic DNA which contained introns.

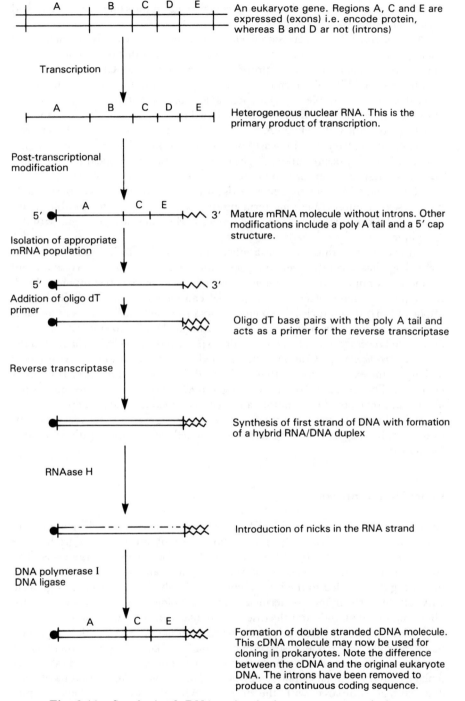

Fig. 2.11 Synthesis of cDNA molecules by reverse transcription.

The cDNA may now be used in cloning experiments in the same way as described earlier for prokaryote DNA.

A problem with the expression of cloned eukaryote enzymes is the inability of bacteria to make the appropriate post-translational modifications. For example, many extracellular eukaryote proteins are glycosylated. Glycosylation may enhance the stability of enzymes and is certainly important when considering clinical uses for example (see Chapter 5 for further details). In an effort to minimize these problems DNA has been cloned into special shuttle vectors which may be used in the well-defined *E. coli* system but can also be transferred to yeasts. As yeasts are eukaryotes, at least some of the appropriate post-translational modifications may occur, although glycosylation is rarely identical to that which occurs in higher organisms. The problem of introducing the correct post-translational modification into enzymes is an area that still requires a considerable basic research input. One long-term possibility is to clone into suitable mammalian cells in culture, although as mentioned earlier there are problems with this technique.

An excellent example of the application of cloning techniques to enzyme technology has recently been published and this concerns the work on calf chymosin. A number of biotechnology companies have used recombinant DNA techniques and have successfully expressed calf chymosin in *E. coli*. This is an important first step but clearly an enteric organism such as *E. coli* is not an ideal source of a food-grade enzyme. However, a major Dutch enzyme company has announced recently that it has successfully expressed the calf chymosin gene in the yeast *Kluyveromyces* sp. Coupled with a single-step purification procedure, the activity of the extract exceeds that of the traditionally obtained preparations of chymosin. The great advantage of cloning into *Kluyveromyces* is that the organism is already approved by the regulatory authorities for food use and so there should be no major hurdle to the acceptance of the cloned enzyme. It is clear that this example is but one of a growing number of cloned enzyme preparations suitable for industrial use.

Concluding remarks

In retrospect it is possible to identify a number of stages in the development of industrial enzymes. Originally, preparations were obtained mainly from crude plant and animal extracts with some contribution from micro-organisms. The combination of limited supply of suitable plant and animal material and increasing demand led to the introduction of microbial enzymes on a large scale with all the advantages of fermentation technology, i.e. no problems with obtaining raw materials and the discovery of many novel activities. Many of the microbial enzymes, particularly those with novel activities, e.g. alkaline proteinases, account for a major share of the total market. However, there are a significant number of microbial preparations that were intended as direct replacements for animal or plant enzymes, e.g. microbial rennets, that have some disadvantage compared with the original. With the advent of gene cloning techniques, it is now possible to compromise and to produce genuine animal or plant enzymes directly from micro-organisms. Although the limited number of acceptable enzyme-producing micro-organisms is unlikely to change significantly, there can be little doubt that techniques such as gene manipulation will allow the production of many novel activities in the future.

Chapter 3

The extraction and purification of enzymes

Introduction

The extraction and purification of an enzyme presents the enzymologist with a paradox. A different purification procedure is usually necessary to isolate each enzyme, yet there are relatively few techniques that are available for the separation of proteins. However, by sequential use of a number of these techniques and by exploitation of the subtleties inherent in the methodology, it has been possible to isolate many enzymes in a pure or nearly pure state. One of the aims of this chapter is to illustrate the essential principles of enzyme purification and to provide some indication of how to measure the degree of purity of the final product (Colowick and Kaplan, 1971; Scopes, 1982).

It is appropriate at this point to ask whether there is a need to purify an enzyme at all. It has become almost axiomatic in biochemistry that if an enzyme is worth using then at the outset it should be purified to homogeneity. However, for many biotechnological purposes this is not necessary and many processes would be uneconomic if the enzyme had to be pure. For most applications it is probably sufficient that any contaminating activities should not affect the substrate(s), the product(s) or the enzyme(s) involved in a particular process. However, there are some applications in the fields of pharmaceutics and clinical medicine where it is

The extraction and purification of enzymes

essential that contaminating proteins are minimized or avoided to prevent unwanted side effects to the patient.

A vast number of purification procedures have been devised for the isolation of proteins and some enzymes have been purified satisfactorily using more than one approach. The process of purifying enzymes, although complex at first sight, can be reduced to the sequential application of a few simple methods (Bonnerjea et al., 1986). First extraction and purification on the laboratory scale will be considered, and then the scale-up of the process will be described.

Enzyme extraction

First and foremost a suitable extract of the enzyme-containing tissue must be made. For many simple hydrolytic enzymes of microbial origin, the protein is secreted into the growth medium and removal of the cells by centrifugation is the only step that is required. This technique works particularly well with Gram positive bacteria such as *Bacillus subtilis* (producer of the proteinase subtilisin) and to a lesser extent with Gram negative organisms such as *Klebsiella pneumoniae* (producer of the α-1-6-glucosidase, pullulanase). However, the presence of an outer membrane in Gram negative organisms can lead to complications and perhaps only partial release of enzyme into the medium.

In some eukaryote micro-organisms, such as the yeast *Kluyveromyces marxianus*, the enzyme may be located outside the cell membrane but is not released into the growth medium. Release of the enzyme often involves partial or total destruction of the cell. For instance, the inulinase (invertase-like enzyme that degrades $\beta 2 \rightarrow 1$ polyfructans, e.g. inulin) of *K. marxianus* can be liberated from the surface of the yeast by autolysis of the cells at 50 °C for 14 hours.

Obtaining intracellular enzymes from micro-organisms can produce greater problems than those indicated above. Techniques for the lysis of cells such as mechanical disruption (e.g. French pressure cell, grinding with alumina), sonication, detergent treatments and osmotic shock are among those that have been used successfully to release intracellular enzymes. However, the enzyme extract is now contaminated with many other components from the cells and this should be recognized when designing subsequent purification steps. In particular, lysed bacterial cells liberate large quantities of nucleic acids, so consequently one of the earliest stages in the purification procedure is designed to remove this contaminant.

The techniques that are used for the extraction of enzymes from animal tissues are, on the whole, similar to those employed for the extraction of intracellular enzymes from microbial eukaryotes (e.g. yeasts). Homogenization of the selected tissue in a suitable buffer is by far the most commonly used technique. Differential centrifugation is often used as a second stage to select the appropriate cellular subfraction or organelle. Isolation of an appropriate organelle can in itself afford a considerable purification of the enzyme, although on a large scale this may not be a practicable approach.

Extraction of enzymes from plant tissues presents a whole new set of problems and the comparatively small number that have been isolated is in part a reflection of the difficulties that are encountered. The cell walls of plants present a formidable challenge to the enzymologist. The forces needed to destroy the cell wall are so great that the desired enzymes are frequently denatured in the process. Furthermore, many plants contain phenolic compounds that are oxidized enzymically (polyphenol oxidases) in the presence of molecular oxygen to give products that can rapidly inactivate many enzymes. Nonetheless, there are a number of examples of useful enzymes that have been obtained from plants by simple means for biotechnological processes, e.g. the proteinases bromelain (pineapple) and papain (papaya).

The pH, ionic strength and composition of the medium that is to be used for the extraction of an enzyme is all important to the success of this and subsequent stages of the purification. There is a range of reagents (Table 3.1) that has been used in extraction buffers to minimize denaturation or degradation of the desired enzyme. The choice of an appropriate combination of reagents is usually found by conducting a series of pilot experiments. Expedience means that almost invariably

Table 3.1 Components used in extraction buffers*

Class of component	Examples	Uses
Thiol reagents	Dithiothreitol	Protection of active sites sulphydryl groups from oxidation
Chelators	EDTA EGTA	Chelators of cations, particularly heavy metals. EGTA is specific for Ca^{2+}
Detergents	Tween 20	Solubilization of membrane-bound proteins or disruption of vesicles
Substrates/substrate analogues	Substrates Competitive inhibitors	Often help to stabilize the enzyme against heat inactivation or extremes of pH
Adsorbants	Polyvinylpyrollidone	Binds reactive compounds useful for extracts from plants
Inhibitors	EDTA Alkylating reagents	Inhibitors of degradative enzymes such as proteinases and certain glycosidases

* It is often essential that an extraction buffer not only has the correct pH and ionic strength but that appropriate stabilizers should be added. A range of chemicals is presented rather than an exhaustive list. Not all of these reagents are universally useful; for example, a metalloproteinase would be inactivated by EDTA whereas urease would be stabilized.

a compromise is reached whereby a satisfactory extraction medium is found, although it may well be that a different combination of reagents could have been more effective.

Enzyme purification

It is likely that the enzyme that you require is present in the extraction medium as only a minute proportion of the total protein (typically, 0.01–1%). Therefore, the early stages of enzyme purification should be designed to remove as much of the unwanted protein as possible in the easiest possible way. Ideally, it should be possible to exploit a particular facet of the stability of the enzyme under some adverse conditions.

Fractionation
Heating the mixture or adjusting the pH to extreme values can be very effective. For example, aspartate aminotransferase may be isolated from pig hearts by briefly heating (20 min) the extraction mixture at 72 °C in the presence of the substrate 2-oxoglutarate and the competitive inhibitor succinate. It is worth noting that if the succinate and the 2-oxoglutarate are omitted at this stage then most of the enzyme activity is lost. The rationale for this increased stability is understood in molecular terms. A molecule of succinate is able to form salt bridges with two arginine residues, effectively introducing a cross-link, whereas 2-oxoglutarate induces the reversible formation of a covalent bond between the coenzyme, pyridoxal phosphate, and the enzyme. Alternatively hyaluronidase, an enzyme of pharmaceutical importance, is effectively extracted from bovine testes using a mixture of acetic acid/hydrochloric acid at pH 2.1. Many of the unwanted proteins, including an important contaminant β-glucuronidase, are denatured and precipitated by the use of this method.

Ammonium sulphate fractionation is an extremely effective early step in the purification of proteins. The technique has the additional advantage of reducing the volume of material, which is often an important consideration at this stage. In principle, the method depends on the ability of high concentrations of ammonium sulphate to bind available water molecules and thus prevent effective solvation of the proteins. For a particular set of conditions each protein will precipitate over a characteristic and reasonably narrow range of ammonium sulphate concentrations. The pH of the medium, the temperature and the concentration of the proteins are important and these factors should be kept as constant as possible between batches if reproducibility is to be obtained. If an ammonium sulphate fractionation at a certain pH is unsatisfactory, then repeating the experiment at a new pH will almost certainly produce results.

The method does have two notable drawbacks. Ammonium sulphate does contain trace quantities of heavy metals that may be sufficient to inactivate an enzyme; this problem may be circumvented easily by the use of good quality (Analar grade) ammonium sulphate. The other disadvantage is that the enzyme preparation contains a high concentration of ammonium sulphate and usually this has to be removed by the use of dialysis, ultrafiltration or desalting columns before proceeding to the next stage of the purification.

An alternative to ammonium sulphate fractionation is to use organic solvents, particularly acetone or alcohols, to fractionate proteins. Solvents should be used only at low temperatures (<0 °C), otherwise most enzymes are rapidly denatured. Also, acetone and alcohols are highly inflammable and their use has resulted in several fatal accidents in laboratories, therefore a temperature lower than the flash point of the solvent is advisable on safety grounds. In general, the use of solvents has lost favour over the years, although there are some particularly successful fractionations based on this methodology, e.g. Cohn's fractionation of blood proteins with ethanol.

For intracellular enzymes of bacterial origin it is often useful to remove the nucleic acids at this stage. The nucleic acids may be precipitated with protamine sulphate or, alternatively, bound to an anion-exchange column (e.g. DEAE–cellulose), leaving the enzyme free in solution. The most noticeable change after the removal of nucleic acids is the considerable reduction in the viscosity of the extract.

Ion-exchange chromatography
The separation of proteins on the basis of charge characteristics by ion-exchange chromatography is one of the most useful methods of purification. Ion exchangers are made from a nominally inert polymer, usually polystyrene or cellulose that has been covalently modified to contain either negatively or positively charged groups (Table 3.2). Cellulose ion exchangers are used in preference for most purifications as this polymer has a lower charge density and tends to have less non-specific interactions with the proteins, thus minimizing inactivation. Ion-exchange resins may be used either in a batch mode or in chromatography columns. Batch mode tends to be used with very crude protein solutions whereas the column method is more usual and is applicable to the resolution of complex mixtures at later stages in the purification.

In general, proteins are applied to a column under conditions that will promote maximum binding and then they are eluted selectively by altering the ionic conditions. Thus, for a cation exchanger, e.g. CM–cellulose, a low ionic strength and pH 5 are ideal conditions for binding proteins onto the resin. This pH ensures that the cation exchanger is negatively charged (typical pK of the carboxymethyl groups $\sim$4) and that the majority of proteins will be below their isoelectric points and therefore will be positively charged, i.e. polycations. The enzymes are eluted selectively from the column by increasing either the ionic strength, or the pH, or by a combination of these two parameters. Generally, it is better to elute the proteins using a continuous rather than a step-wise gradient as this reduces the possibility of artefacts due to frontal elution effects.

For anion-exchange chromatography the crude protein mixture is applied to the column at low ionic strength and at approximately pH 8 (DEAE–cellulose has a pK of 9.5). Enzymes are eluted sequentially by either an increase in ionic strength, a decrease in pH, or by a combination of both methods. Ion-exchange chromatography has the added advantage that nucleic acids may be bound to DEAE–cellulose and thus removed from the enzyme solution.

Table 3.2 Structure of substituents on various cellulose ion exchangers

Substitution	Structure	Approximate pK value(s)
Anion-exchangers		
Diethylaminoethyl (DEAE)	R—O(CH$_2$)$_2$—$\overset{+}{\text{NH}}$(CH$_2$CH$_3$)$_2$	9.5
Triethylaminoethyl (TEAE)	R—O(CH$_2$)$_2$—$\overset{+}{\text{N}}$(CH$_2$CH$_3$)$_3$	9.5
Diethyl 2-hydroxypropylamino (QAE)	R—O(CH$_2$)$_2$—$\overset{+}{\text{N}}$(CH$_2$CH$_3$)$_2$CH$_2$CHOHCH$_3$	Strongly basic
ECTEOLA	Undefined mixed amines	7.5
Cation-exchangers		
Carboxymethyl (CM)	R—CH$_2$COO$^-$	4.0
Phosphoryl (P)	R—O—P(=O)(O$^-$)(O$^-$)	1.5, 6.0

Ion exchange is often one of the most successful steps in a purification scheme, giving a high increase in specific activity combined with a good recovery of total enzyme activity. However, to use ion exchange to full advantage it is important that the crude enzyme solution is added to the column in a low ionic strength buffer and preferably in the buffer that has been used to equilibrate the column.

Chromatofocusing

A useful and particularly powerful variant of ion-exchange chromatography is the technique of chromatofocusing. Proteins are eluted from an ion-exchange column using a carefully selected buffer, and as the name of the method implies a focusing effect occurs, hopefully concentrating the enzyme of interest. In conventional ion-exchange chromatography the column is eluted initially with the equilibration buffer followed by an appropriate, gradual adjustment of the ionic conditions. With chromatofocusing the column is equilibrated using one buffer and then, after application of the sample, a second buffer of different pH is used for elution. Providing that the composition and strength of the eluting buffer are compatible with the exchange capacity of the column a linear pH gradient will be generated *in situ* as a result of the buffering capacity of the resin. This self-generating pH gradient has a focusing effect and molecules with a particular isoelectric point are concentrated together. The technique produces high resolution separations but has the disadvantage of being rather too expensive for most large-scale operations. However, it is true to say that because of the focusing effect much higher loading of columns may be allowed compared with conventional ion exchange.

Gel permeation chromatography

Molecular size separation is often involved as one of the final stages in the purification of an enzyme. This technique is not usually used early on in the purification process because of the limitations on sample size that are inherent in the method. Beads of cross-linked polysaccharides such as Sephadex (dextran based) and Sepharose (agarose based) are the most commonly used materials for gel permeation. The method is dependent on the ability of a particular molecule to enter the pores in the gel beads. Compounds of large molecular size will not be able to penetrate the pores and will be eluted quickly in the non-included mobile phase (usually less than 40% of the total column volume). Smaller molecules will be able to enter the beads, have access to the stationary phase, and will take longer to be eluted from the column. Therefore, protein molecules are eluted from the column in order of decreasing molecular size. Providing that the protein molecules are approximately spherical in shape, then the order of elution can be related directly to the molecular weight. With care this methodology can afford a useful purification step often with a 100% yield of enzyme activity.

Although gel filtration using Sephadex or related materials is an extremely successful technique, there are a number of disadvantages to be considered. The amount of sample that can be applied to the gel is generally restricted to 1–2% of the total column volume; with fairly simple separations this proportion may be increased. Sephadex and Sepharose tend to be compressible and so care must be taken not to exceed the recommended maximum hydrostatic pressure in an

The extraction and purification of enzymes

attempt to increase the flow rate. The gels also tend to retain a residual negative charge and so a buffer with a minimum ionic strength (mol l^{-1}) of 0.02 is advisable to prevent ion-exchange effects.

To overcome the disadvantages of compressibility a number of alternatives to the cross-linked polysaccharide gels have been manufactured. These include vinyl polymers (Fractogel), controlled pore glass and copolymers of allyldextran and bisacrylamide (Sephacryl).

Affinity chromatography
A more recent development in protein purification has been the advent of affinity chromatography (Dean *et al.*, 1985). This method utilizes the specific interaction between an enzyme and a suitable immobilized ligand such as a substrate, product, coenzyme, inhibitor or some other compound, e.g. certain organic dyes. The original application of this methodology, which was reported as long ago as 1910, was the selective adsorption of amylase onto insoluble starch. However, the technique has developed fully in recent years largely as a result of the introduction of suitable matrices and chemical methodology for the immobilization of the ligands.

With affinity chromatography it should be possible, under ideal circumstances, to pass a crude protein solution down a ligand-containing column so that only the enzyme of interest will be retained. Subsequently, the bound enzyme may be eluted using low concentrations of substrate, coenzymes or some other suitable molecule. The degree to which this ideal is achieved will vary enormously depending on the enzyme and the immobilized ligand that is used. Affinity chromatography will work best if the dissociation constant for the interaction in free solution between the ligand and the enzyme is of the order of 10^{-4} to 10^{-8} mol l^{-1}. Chromatography based on components with more extreme values for the dissociation constant result in either poor adsorption of the enzyme to the column or alternatively an essentially irreversible binding.

The number of possible permutations of ligands and enzymes suitable for affinity chromatography is vast. Naturally, the more specific the interaction, the more successful the technique is likely to be. One of the disadvantages of having such a range of interactions is the difficulty in producing commercial affinity chromatography matrices that are suitable for a wide range of applications. Also, many of the ligands, particularly some of the coenzymes, are too expensive for large-scale use. However, a number of group-specific affinity chromatography matrices have been developed and used with remarkable success. For instance, the triazine dye Cibacron Blue F3G-A has been used as a ligand for the isolation of enzymes containing an adenine nucleotide binding site, e.g. dehydrogenases and kinases. The original discovery that this dye was specific for certain enzymes was totally accidental but with hindsight it is now possible to rationalize the observation on the basis of structural similarities between NAD^+ and Cibacron Blue F3G-A (Fig. 3.1). Some spectacular one-step purifications have been achieved using Cibacron Blue F3G-A bound to carriers (e.g. Blue-Sepharose, Matrex gel Blue A). Yeast glucose-6-phosphate dehydrogenase was purified 4460 times with a 60% recovery of total activity using this method. Since the discovery

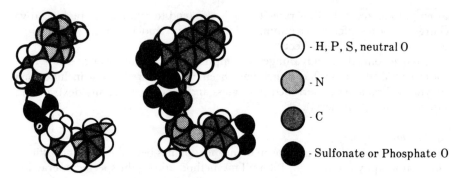

Fig. 3.1 Corey–Pauling–Koltin structural model of NAD^+ (*left*) and Cibacron Blue F3G-A (*right*) (©Amicon Corporation, 1980).

of Cibacron Blue F3G-A as an affinity-ligand, numerous other dyes have been screened for similar properties and there is now a range of these products available from various manufacturers. However, it should be noted that, on the whole, these general affinity chromatography methods will be far less specific than methods designed with a particular application in mind.

A variation of affinity chromatography is the fractionation of enzymes on immunoadsorbents. Providing that sufficient antibodies can be raised against the enzyme, then these may be attached to a matrix. The strong affinity between antibody and antigen (the enzyme in this case) will allow the selective binding of the required enzyme to occur. However, subsequent desorption can present a problem. Often fairly vigorous conditions are needed to elute the enzyme, and the risk of denaturation is ever present, unless there is a specific molecule that is able to compete for the antibody binding site. Typically, desorption is achieved by adjustment of the pH, by reducing the polarity of the buffer (e.g. addition of dioxan), by the use of dissociating agents (e.g. urea) or the use of chaotropic ions (e.g. SCN^-). Alternatively, the enzyme may be removed electrophoretically. Provision of sufficient quantities of antibody in reproducible batches has always been a limitation of this technique. However, with the advent of monoclonal antibodies the potential for using this method is much greater (Chase, 1984). Nonetheless, this is an expensive route to enzyme purification and applications have been rather limited.

High performance liquid chromatography
A problem with all of the chromatographic techniques is that they are time-consuming and usually need to be done at 4 °C to reduce the possibility of enzyme inactivation or microbial contamination. The recent development of new column-packing materials for high performance liquid chromatography has expanded the use of this technique to the purification of proteins (Waterfield, 1986). Column packings have been developed for ion exchange, chromatofocusing, gel permeation and affinity chromatography. Using standard columns approximately 25 mg of protein can be fractionated at room temperature in less than 1 hour, thus replacing chromatographic steps that often required overnight running. Although the application of high performance liquid chromatography to the separation of

The extraction and purification of enzymes

proteins is still at the developmental stage, there is little doubt that this technique will emerge as an important tool for enzymologists. Currently, systems are being developed that are capable of handling sample volumes of 10–20 litres and containing many grams of protein (Dwyer, 1984).

Purification of an enzyme will involve the successive use of a number of techniques and a typical example is presented in Table 3.3.

Large-scale Purification

Most of the principles that have been described for the isolation and purification of enzymes can be adapted to large-scale operation although the detailed methodo-

Table 3.3 Comparison of purification methods for β-galactosidase obtained from *Escherichia coli*

Method 1 (published 1965)	*Method 2* (published 1971)
E. coli cells	*E. coli* cells
Harvest by centrifugation	Harvest by centrifugation
Disrupt by sonication	Disrupt by sonication
$(NH_4)_2SO_4$ fractionation	$(NH_4)_2SO_4$ fractionation
Chromatography on Sephadex G-200	Affinity chromatography on Agarose-p-aminophenyl β-D-thiogalactosidase
$(NH_4)_2SO_4$ precipitation	
Chromatography on DEAE Sephadex A-50	

Method 3 (published 1978)

E. coli cells

Harvest by centrifugation

Disrupt by high pressure homogenization

Heat treatment

$(NH_4)_2SO_4$ precipitation

Approximate yields and activities

Method	1	2	3
Mode	Batch	Batch	Semi-continuous
Specific activity (units mg^{-1})	340	320	132
Yields	3 g	0.18 g	25 g hour^{-1}

logy may well be quite different (Dunnill, 1983). Engineering problems such as fluid flow, mass transfer and heat transfer, that are of no consequence in the laboratory, become of paramount importance in the scale-up of a process. In effect this often means that the number of different options available to the enzymologist is more limited.

The release of intracellular enzymes by mechanical means is restricted to the use of ball mills or homogenizers. The former works by rapidly agitating the tissue in the presence of glass balls resulting in the disruption of cells and the release of enzymes. The latter depends on the cutting action of rapidly rotating blades.

Centrifugation of materials also present a problem on the large scale. High-speed centrifugation is largely impractical as the high gravitational force involved would make industrial scale equipment hazardous to use. Therefore, the technique is really best suited to the removal of large particulate material and a number of different designs of centrifuge are available for this purpose, e.g. tubular bowl and disc bowl centrifuges. An alternative to centrifugation is to filter samples through a bed of filter aid such as Celite.

Dialysis and the concentration of dilute enzyme solutions are best performed using an ultrafiltration apparatus (Kroner *et al.*, 1984b). Also, this technique allows a crude separation of proteins on the basis of molecular weight, dependent on the choice of membrane. The technology is well tried and tested, although there may be problems of minimizing concentration polarization. This phenomenon is the accumulation of high concentrations of protein on the surface of the membrane which causes a reduced flow-rate through the system.

A method that may prove to be useful for the large-scale recovery and purification of enzymes is liquid extraction in aqueous-phase systems (Kroner *et al.*, 1984a). Typically, two-phase systems can be produced using various combinations of either polyethylene glycol/potassium phosphate or polyethylene glycol/crude dextran. The technique enables cell debris, contaminating proteins, nucleic acids and polysaccharides to be removed without having to use centrifugation or filtration equipment. In some cases aqueous-phase extraction may replace chromatographic steps such as gel permeation or ion exchange, but the enzyme would still require to be concentrated after this step. Also, the cost of dextrans may be prohibitively expensive.

One might expect scale-up of the various chromatographic procedures such as ion exchange and gel permeation chromatography to be relatively straightforward. Although some successful examples have been documented, scale-up often presents many problems, not least of which is the maintenance of a uniform fluid flow through the column. However, several manufacturers of chromatographic media provide a technical service and the appropriate equipment to scale-up for most applications. An alternative to column chromatography is to consider batch processes in stirred tanks.

It is difficult to generalize about the methods of scale-up as each process will present its own problems. However, most of the techniques available for the purification of proteins may be adapted in some way or other to large-scale production. The data presented in Table 3.3 illustrate a procedure for the purification of an enzyme on a large scale.

Enzyme specification

Enzymes that have been produced for commercial use will have to conform to certain specifications. In particular, there are rigorous regulations governing the source and use of enzymes within the food and pharmaceutical industries. Furthermore, each country has its own set of regulations, many of which are incompatible with those of other nations.

For relatively crude preparations of enzymes it may be sufficient to specify the source and to guarantee a minimum specific activity. In addition, it is often helpful to know the activities of any contaminating proteins that may have a detrimental effect on the intended application. Naturally, most manufacturers of enzymes provide considerably more data than this minimum specification.

The use of enzymes for pharmaceutical purposes requires that a more comprehensive specification than that outlined above be provided. For instance, if an enzyme is to be injected intravenously into a patient, then it is essential that none of the contaminants are toxic. Therefore, the regulatory authorities will require that the enzyme not only has a minimum specific activity but also complies with a whole series of other tests. As a minimum, an enzyme that was to be intravenously administered would have to be bacteriologically sterile and free of both pyrogens and of proteins that may cause anaphylactic shock, e.g. albumin.

Should it be necessary to obtain a 'pure' enzyme, then there are a number of analytical methods that are available that help to ascertain whether contaminating proteins are present. Most commonly, the presence of a single protein band after electrophoresis on sodium dodecyl sulphate-containing polyacrylamide gels under reducing conditions is indicative of a pure enzyme. If an antibody is available, then the techniques may be extended by the use of Western blotting (Fig. 3.2) to show that the enzyme is coincident with the protein band. Isoelectric focusing is also a discriminating test of enzyme purity. However, it is impossible to prove absolutely that an enzyme is 100% pure and these analytical techniques should not be used with blind faith.

The specification of an enzyme will be dependent on both the process requirements and the legal constraints. It should be noted that there is little point in developing an exotic process unless there is a reasonable chance that the regulatory authorities will sanction the use of the intended enzyme preparation. Even if this is the case then the cost of gaining regulatory approval, often in a series of countries each of which has its own codes of practice, can be prohibitively expensive and ultimately may be the deciding factor in the economics of developing a process.

Concluding remarks

Enzyme purification is essentially a practical process worked out largely by trial and error. Although educated guesses and the judicious combination of different techniques will help, there is no substitute for actual experimentation. All purification procedures are a compromise between total yield of enzyme and

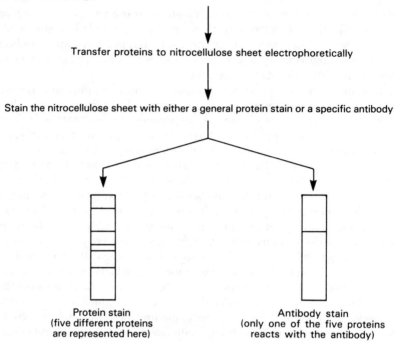

Fig. 3.2 The principles of Western blotting.

obtaining the highest specific activity; for most biotechnological applications the former criterion is of greater importance. Maintaining the stability of the enzyme during the purification procedure is of paramount importance and careful handling of the material and low temperature can do much to improve yields.

Chapter 4

Kinetic properties and reactor design

General considerations

Kinetic studies represent just one of a range of tools available to the biochemist investigating the structure and function of enzymes, and a range of complex mechanistic models have been developed (Fersht, 1977). However, when considering the technological application of enzymes our goals are more clearly defined and it is often sufficient to restrict ourselves to an examination of the effects of the action of an enzyme rather than a detailed kinetic study. When considering the commercial suitability of an enzyme catalyst there are three major criteria that must be assessed:

(1) the rate of reaction (catalytic activity)
(2) the extent of reaction (equilibrium constant)
(3) the duration of usable activity (stability)

The relative importance of these criteria will vary with the application envisaged. For example, a sensor would require a stable enzyme but the extent of reaction may not be critical, whereas for a production system all three factors must be considered. In all cases cost will be a significant factor.

Rate of reaction

Enzymes, as catalysts, promote chemical reactions but are not irreversibly modified themselves. Rate enhancements are attributed to a lowering of the activation energy of the reaction. In simple terms this is explained by the reactant physically binding to the enzyme at an 'active site', forming a metastable complex (Palmer, 1981). The activation energy for the breakdown of this complex is then considerably lower than that for the breakdown of the reactant alone. Starting from the simple concept we can propose a reaction scheme as follows:

$$\text{Enzyme} + \text{Reactant} \rightleftharpoons \text{Complex} \longrightarrow \text{Enzyme} + \text{Product}$$
$$[E] \quad\quad [R] \quad\quad\quad [ER] \quad\quad\quad [E] \quad\quad [P]$$

The complex formation step is regarded as being reversible as the complex has a similar energy state to the enzyme–reactant mixture. The subsequent breakdown of the complex to give products will be exothermic and this stage may be effectively irreversible in many cases. The rate expression for this type of reaction is usually derived using a steady state assumption, i.e. the complex concentration remains constant.

$$[E] + [R] \underset{k_{-1}}{\overset{k_1}{\rightleftharpoons}} [ER] \overset{k_2}{\longrightarrow} [E] + [P]$$

Making a steady state assumption,

$$\overset{\text{Formation}\quad\text{Breakdown}}{\frac{d[ER]}{dt} = k_1 [E][R] - (k_{-1} + k_2)[ER] = 0}$$

Therefore,

$$k_1[E][R] = (k_{-1} + k_2)[ER]$$

For this assumption to be accurate, the rate of reaction must be linear and so is usually determined as close to time zero as possible, before the reactant concentration changes appreciably (the initial rate).

The above equation can be rearranged to give:

$$\frac{[E][R]}{[ER]} = \frac{(k_{-1} + k_2)}{k_1} = K_m \quad\quad (1)$$

where K_m is the Michaelis constant.

The law of mass conversion requires that the total enzyme ($[E_0]$) does not change.

$$[E_0] = [ER] + [E]$$

Therefore,

$$[E] = [E_0] - [ER] \quad\quad (2)$$

Kinetic properties and reactor design

Substitution for $[E]$ in equation (1) and rearranging gives

$$[ER] = \frac{[E_o][R]}{K_m + [R]} \quad (3)$$

The dimensions of this expression are in units of concentration. The rate constant k_2 is a first-order rate constant having units of time^{-1}. Multiplication of both sides of equation (3) by k_2 will give the rate of product concentration with time as a function of the maximum possible rate, the Michaelis constant and the reactant concentration.

$$\frac{dP}{dt} = k_2[ER] = \frac{k_2[E_0][R]}{K_m + [R]}$$

This is usually written in the form of

$$v = \frac{V_{max}[R]}{K_m + [R]} \quad \text{the Michaelis–Menten equation} \quad (4)$$

To predict the performance of an enzyme-utilizing system we need to determine V_{max} and K_m for the enzyme of interest. This can be achieved by measuring the rate of reaction at a range of reactant concentrations (Fig. 4.1). It can be shown algebraically that a rate equal to half V_{max} is obtained when the reactant concentration equals K_m. It can be seen from equation (4) that at values of $[R]$ much lower than K_m the expression can be simplified to

$$v = \frac{V_{max}[R]}{K_m} \quad \text{i.e. first order, as } K_m + [R] \simeq K_m$$

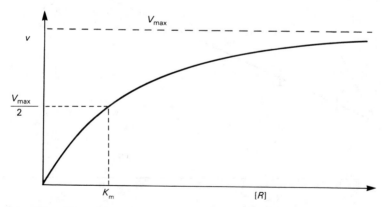

Fig. 4.1 The effect of reactant concentration on the rate of an enzyme-catalysed reaction.

At values of [R] much higher than K_m the value of equation (4) approaches

$$v = \frac{V_{max}[R]}{[R]} \quad \text{i.e. zero order}$$

Determination of kinetic constants

The hyperbolic nature of the v against [R] curve makes the determination of constants difficult from this simple plot. Invariably there is a tendency to underestimate the value of V_{max}; however, it serves to demonstrate the importance of using a range of concentrations both higher and lower than the K_m value, to avoid obtaining only first-order or zero-order data. The most commonly used graphical method for determining kinetic constants is the double reciprocal plot, where reciprocals are taken on each side of equation (4) to give

$$\frac{1}{v} = \frac{K_m}{V_{max}} \cdot \frac{1}{[R]} + \frac{1}{V_{max}}$$

Plotting $1/v$ against $1/[R]$ gives a straight line relationship, allowing both constants to be determined by extrapolation. However, by taking reciprocals the most significance is placed on the rates obtained at low reactant concentrations which will be subject to the greatest experimental error (Fig. 4.2) (Eisenthal and Wharton, 1981).

Several alternative plots have been described in the literature but most suffer

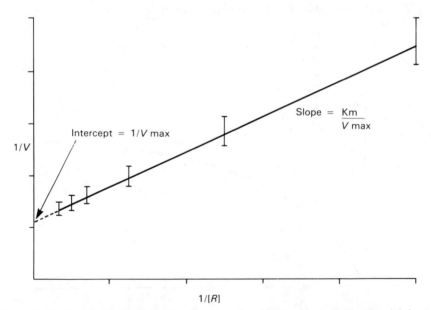

Fig. 4.2 A double reciprocal plot showing the effects of a $\pm 10\%$ error in the v value.

Kinetic properties and reactor design

to some extent by imparting bias to the results. One of the more recent approaches which appears to avoid these drawbacks is the direct linear plot (Eisenthal and Cornish-Bowden, 1974). This method is based on treating V_{max} and K_m as variables and $[R]$ and v as experimentally determined constants.

Equation (4) can be rearranged to give

$$V_{max} = \frac{v_1 \cdot K_m}{[R_1]} + v_1$$

for any individual observations $[R_1]$, v_1. This expression will be true for any pair of $[R]$ and v values. Hence,

$$\frac{v_1 \cdot K_m}{[R_1]} + v_1 = \frac{v_2 \cdot K_m}{[R_2]} + v_2$$

Solving for K_m

$$K_m = \frac{(v_2 - v_1)}{\left(\frac{v_1}{[R_1]} - \frac{v_2}{[R_2]}\right)}$$

A similar approach can be used for V_{max}.

$$K_m = V_{max}\frac{[R_1]}{v_1} - [R_1] = V_{max}\frac{[R_2]}{v_2} - [R_2]$$

Solving for V_{max} gives

$$V_{max} = \frac{([R_2] - [R_1])}{\left(\frac{[R_2]}{v_2} - \frac{[R_1]}{v_1}\right)}$$

Values of V_{max} and K_m can be determined for every pair of observed values and the median values ascertained. This can be carried out using a computer or alternatively a simple plot can be constructed (Fig. 4.3). A line is drawn connecting $-[R_1]$ to v_1 and extending into the first quadrant. This process is then repeated for the rest of the data, i.e. $-[R_2]$, v_2; $-[R_3]$, v_3. If the data are error-free, then a unique intersection point will be obtained from which the values of K_m and V_{max} may be read directly. Usually, more than one intersection point will be obtained and it is crucial to the method that the median value is taken as the best estimate of K_m and V_{max}. The number of potential intersections is given by the following equation:

$$\text{Number of intersections} = (n/2) \times (n-1)$$

where n = number of observations.

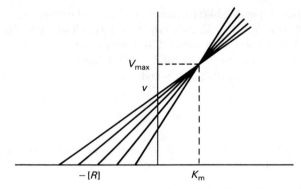

Fig. 4.3 Graphical representation of the direct linear plot.

The single reactant irreversible reaction represents the simplest case of enzyme kinetics. However, in practice, the reaction may be affected by the presence of inhibitors which can combine with the enzyme and form a non-active complex. In many technological applications it may be possible to avoid problems of enzyme inhibition by restricting the number of compounds present in the feed streams. The two examples of inhibition which cannot be avoided in this way are those caused by high concentrations of either the reactant or product, phenomena peculiar to enzymes. In the first case, elevated concentrations of reactant lead to a second molecule binding to the enzyme.

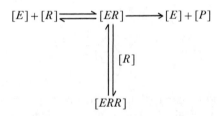

This is termed uncompetitive inhibition and leads to a rate expression of the form

$$v = \frac{V_{max}[R]}{[R]\left(\dfrac{1+[R]}{K_i}\right) + K_m} \qquad (5)$$

where $K_i = [ER][R]/[ERR]$.

With product inhibition it is more common for the inhibitor molecule to be sufficiently similar in structure to the reactant for it to compete for the active site of the enzyme.

Kinetic properties and reactor design

$$[E]+[R] \rightleftharpoons [ER] \longrightarrow [E]+[P]$$

$$\Updownarrow [P]$$

$$[EP]$$

This leads to an expression of the form

$$v = \frac{V_{\max}[R]}{[R] + K_m \left(\dfrac{1 + [P]}{K_i}\right)} \qquad (6)$$

where $K_i = [E][P]/[EP]$.

In the case of reactant inhibition the effects can be seen clearly from a simple plot of v against $[R]$ (Fig. 4.4).

For product inhibition the rate of reaction must be studied in the presence of a significant product concentration. The effects of product concentration would not normally be seen in initial rate studies as there is no appreciable product present. In designing an assay for enzyme activity it is important to cover the concentration range about the K_m value. With a new enzyme of unknown properties it may be necessary to find this range by trial and error. Once a crude estimate has been obtained for the value of K_m, the usual range to use would be in the region 0.5–5 K_m. The lower value would be dictated by the accuracy of measurement of

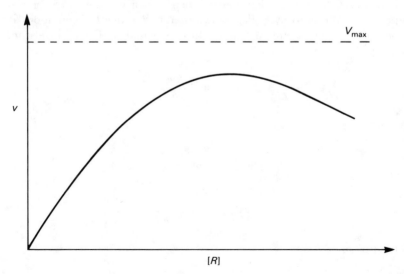

Fig. 4.4 Effect of reactant inhibition on the rate of an enzyme-catalysed reaction.

reactant and product concentrations. In this first-order region small errors in $[R]$ will have a large effect on the value of v that is obtained. The higher value would be determined by reagent costs and the problems of non-specific inhibition. For each case of inhibition values of K_i can be calculated from the apparent values of V_{max} and K_m obtained from a series of experiments at different inhibitor concentrations. The relationship between apparent constants and inhibitor concentrations can be used to determine the true values of the constants (Palmer, 1981).

Effects of temperature and pH
The Michaelis–Menten equation shows that the rate of an enzyme-catalysed reaction is

$$k_2 [ER]$$

So the rate of reaction will be dictated by the concentration of $[ER]$ and the value of k_2. In a simplified assessment, the variation of the rate of an enzyme-catalysed reaction with temperature can be described in terms of a change in the value of k_2. As with simple chemical reactions, this can be described by the Arrhenius relationship (Morris, 1974).

$$k_2 = A \exp(-E/RT) \qquad (7)$$

where A = the Arrhenius constant
E = the activation energy
R = the gas constant
T = temperature

If A and E have been determined, then k_2 can be calculated for a given temperature. Alternatively, if V_{max} is measured under identical conditions at two temperatures, the constants A and E can be calculated. Taking logarithms on equation (7)

$$\ln k_2 = \ln A - \frac{E}{RT}$$

Therefore

$$\ln V_{max^2} = \ln A - \frac{E}{RT^2}$$

$$\ln V_{max^1} = \ln A - \frac{E}{RT^1}$$

Subtracting gives

$$\ln \frac{V_{max}^2}{V_{max}^1} = \ln \frac{k_2^2 [E_0]}{k_2^1 [E_0]} = -\frac{E}{R}\left(\frac{T^1 - T^2}{T^1 T^2}\right) \qquad (8)$$

Kinetic properties and reactor design

In practice it would be more accurate to carry out the experiment at a range of temperatures and plot ln (V_{max}) versus $1/T$. In the case of enzymes the Arrhenius relationship will hold only over a restricted range of temperatures. The limited thermal stability of enzymes results in inactivation at elevated temperatures and a graph of V_{max} versus an extended temperature range would be of the form shown in Fig. 4.5. It must be stressed that the shape of the graph does not imply the existence of a unique temperature optimum. The choice of operating conditions will always result from a balancing of the desirability of a high rate with the need for acceptable stability.

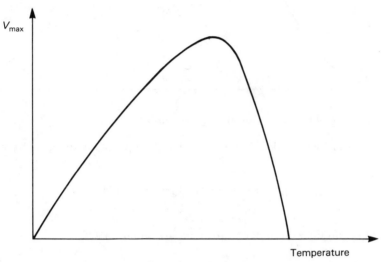

Fig. 4.5 Effect of temperature on enzyme activity (this response might typically be expected over the range 0–100 °C.

While the simplified assessment of the effect of temperature on V_{max} presented here is useful in many cases, examination of equation (1) shows that there is a relationship between K_m and k_2. From this expression it is clear that for enzymes where the magnitude of k_{-1} approaches that of k_2, K_m will also show a temperature dependence.

The pH effects on enzyme activity stem from a requirement for critical groups at the active site to be in the correct ionization state for the reaction to proceed. A similar requirement may apply to reactant ionization. A graph of V_{max} versus pH would usually be carried out to ascertain the pH optimum of the enzyme (Fig. 4.6). The Michaelis constant, K_m, and therefore the initial velocity, would not necessarily show the same variation with pH.

Extent of reaction

In practice not all enzyme-catalysed reactions will be irreversible and so in some cases we must determine the equilibrium point of the reaction. Considering the

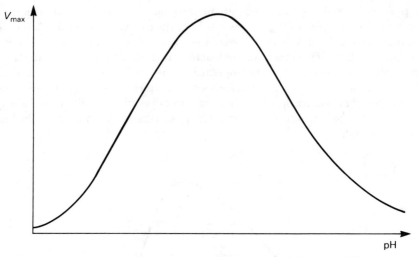

Fig. 4.6 Relationship between enzyme activity and pH.

case of a reversible reaction with intermediate complex formation, we have

$$[E]+[R] \underset{k_{-1}}{\overset{k_1}{\rightleftharpoons}} [ER] \underset{k_{-2}}{\overset{k_2}{\rightleftharpoons}} [E]+[P]$$

At equilibrium the rates in both directions are equal, therefore

$$k_1 [E][R] = k_{-1} [ER]$$

$$k_{-2} [E][P] = k_2 [ER]$$

Rearranging

$$\frac{[ER]}{[E]} = \frac{k_1}{k_{-1}}[R] = \frac{k_{-2}}{k_2}[P]$$

The equilibrium constant $K_{eq} =$

$$\frac{[P]}{[R]} = \frac{k_1 k_2}{k_{-1} k_{-2}}$$

Now, as the individual rate constants cannot be easily determined, the equilibrium constant must be expressed as a function of V_{max} and K_m for both directions.

Forward reaction Backward reaction
$V_{max}(f) = k_2 [E_0]$ $V_{max}(b) = k_{-1} [E_0]$

Kinetic properties and reactor design

Therefore

$$\frac{V_{max}(f)}{V_{max}(b)} = \frac{k_2 [E_0]}{k_{-1} [E_0]}$$

As previously

$$K_m(f) = \frac{k_{-1}+k_2}{k_1}$$

Similarly

$$K_m(b) = \frac{k_{-1}+k_2}{k_{-2}}$$

Therefore

$$\frac{K_m(b)}{K_m(f)} = \frac{(k_2+k_{-1}) \cdot k_1}{k_{-2} \cdot (k_2+k_{-1})}$$

Therefore

$$K_{eq} = \frac{V_{max}(f) \; K_m(b)}{V_{max}(b) \; K_m(f)} \quad \text{the Haldane relationship} \tag{9}$$

So K_m and V_{max} must be determined for both directions. If initial rate measurements are taken, then the reactions can be considered as being irreversible as there will be zero product concentration.

Aspects of enzyme reactor design

Batch processes
Having considered the kinetics of enzymes in free solution, their implications on the design of enzyme reactors can be studied. In the simplest sense a spectrophotometer cuvette can be considered as a batch enzyme reactor. If the change in product concentration or the corresponding change in reactant concentration is monitored, it will tend to follow the relationship described by the integrated Michaelis–Menten equation (Cornish-Bowden, 1979):

$$\frac{-d[R]}{dt} = \frac{V_{max}[R]}{K_m+[R]} = \frac{V_{max}}{\frac{K_m}{[R]}+1} \tag{10}$$

$$\int_{R=R_0}^{R=R} \left(\frac{K_m}{[R]}+1\right) d[R] = \int V_{max} \, dt$$

$$-{}_{R_0}^{R}[K_m \ln\{[R]\}+[R]] = V_{max} t$$

$$(K_m \ln\{[R_0]\} + [R_0]) - (K_m \ln\{[R]\} + [R]) = V_{max}t$$

$$K_m \ln\left\{\frac{[R_0]}{[R]}\right\} + ([R_0] - [R]) = V_{max}t \qquad (11)$$

As $[R_0]$ is the reactant concentration at time zero, $([R_0] - [R])$ will equal the product concentration at any time t, assuming that the reaction stoichiometry is 1:1. So by using equation (11), product concentration can be expressed as a function of time. The enzyme concentration used is expressed in the V_{max} term, i.e. $k_2 . E_0$, the initial enzyme concentration. Theoretically this expression could be used to calculate the K_m and V_{max} of an enzyme from a single product vs time curve. However, the underlying assumptions make this an unreliable method in practice. The assumptions are:

(1) That the enzyme is stable over the time course of the reaction
(2) No product or reactant inhibition is occurring
(3) The reaction is irreversible

For a biochemist studying a new enzyme these would be unwarranted assumptions. But for an enzyme process using a well-characterized enzyme of high stability, the integrated equation can be used to estimate the amount of enzyme and the volume of reactor needed, and also the time taken for a batch process to produce a given quantity of product.

Continuous processes

If the enzyme is immobilized then it is possible to run an enzyme reactor continuously. There are two general models of continuous operation (Vieth *et al.*, 1979).

Continuous Stirred Tank Reactor (CSTR). In this case the reactor vessel is assumed to be perfectly mixed such that the concentration of components in the outlet stream is equal to that in the bulk of the vessel (Fig. 4.7). The effects of the input, output and conversion can be considered in terms of a mass balance on the material entering, accumulating in, and leaving the reactor.

$$\underset{\text{Accumulation}}{V\frac{d[R]}{dt}} = \underset{\text{Input}}{Q[R_0]} - \underset{\text{Removal}}{Q[R]} - \underset{\text{Conversion}}{vV} \qquad (12)$$

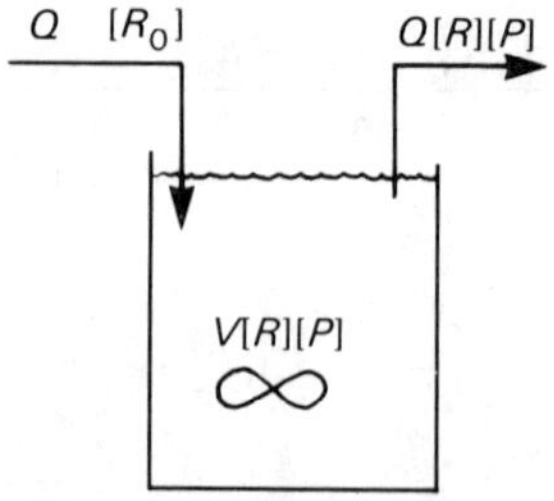

Fig. 4.7 Schematic representation of a continuous stirred tank reactor (CSTR).

Kinetic properties and reactor design

where V = reactor volume (m³)
Q = volumetric flow rate (m³ s⁻¹)
$[R_0]$ = reactant concentration in the feed (kg mol m⁻³)
$[R]$ = reactant concentration in the reactor (kg mol m⁻³)
v = rate of reaction (kg mol m⁻³ s⁻¹)

Dividing equation (12) by V and rearranging gives

$$\frac{d[R]}{dt} = D([R_0]-[R]) - v$$

where $D = Q/V$ = dilution rate (units of reciprocal time).

In practice a continuous reactor would be run under conditions of steady state, i.e. no accumulation of reactant or product in the reactor; hence the accumulation term becomes zero.

$$v = D([R_0]-[R])$$

The expression used to describe v depends on the kinetics displayed by the enzyme of interest. If simple Michaelis–Menten kinetics are assumed, then

$$\frac{V_{max}[R]}{K_m + [R]} = D([R_0]-[R]) \qquad (13)$$

The parameters which can be controlled are D and $([R_0])$, with the ratio of V_{max}/D being the critical parameter. Rearranging equation (13) gives

$$\frac{V_{max}}{D} = \frac{(K_m + [R])([R_0]-[R])}{[R]}$$

It is usual to express this in terms of fractional conversion rather than $[R]$, to allow the formulation of a dimensionless performance equation.

$$X = \frac{([R_0]-[R])}{[R_0]}$$

Therefore

$$([R_0]-[R]) = [R_0]X \text{ and } [R] = [R_0]-[R_0]X$$

Expanding equation (13) gives

$$\frac{V_{max}}{D} = K_m \frac{([R_0]-[R])}{[R]} + [R_0]\frac{([R_0]-[R])}{[R]}$$

Substituting in terms of X

$$\frac{V_{max}}{D} = K_m \frac{([R_0]X)}{([R_0]-[R_0]X)} + [R_0]X$$

Rearranging,

$$\frac{V_{max}}{D} = K_m \left(\frac{X}{1-X}\right) + [R_0]X \tag{14}$$

As discussed in Chapter 6, immobilization will modify the kinetics of enzyme preparations such that both K_m and V_{max} may change. As these changes are difficult to predict, the apparent constants would usually be determined experimentally by conducting a series of experiments at different feed reactant concentrations and measuring the resulting fractional conversion.
Rearranging equation (14),

$$[R_0]X = -K'_m \left(\frac{X}{1-X}\right) + \frac{V'_{max}}{D} \tag{15}$$

So plotting experimentally obtaining values of $[R_0]X$ against $(X/(1-X))$ will give a graph having a slope of $-K'_m$ and an intercept of V'_{max}/D.

Plug Flow Reactor (PFR). In this case it is assumed that no axial mixing occurs in the vessel and that liquid passes through the reactor as a discrete fluid element (Fig. 4.8). In this case the reactor can be modelled using the integrated Michaelis–Menten equation and replacing reaction time with residence time (V/Q), the time which each fluid element spends in the reactor.

Hence, from equation (11),

$$\frac{V'_{max} V}{Q} = K'_m \ln\left\{\frac{[R_0]}{[R]}\right\} + ([R_0] - [R]) \tag{16}$$

V is defined as the total reactor volume. However, a plug flow reactor will often be constructed as a packed bed of immobilized enzyme and so the immobilization matrix may occupy a significant fraction of this volume. As residence time depends on the time spent by the liquid in the reactor, it is a function of the liquid volume (V_1) rather than the total volume (V_{tot}). The voidage of the column can be expressed as

$$\varepsilon = V_1/V_{tot}, \quad \text{so } V_1 = \varepsilon \cdot V_{tot}$$

Therefore the revised form of equation (16) becomes

$$\frac{V'_{max} \cdot V_{tot}\varepsilon}{Q} = K'_m \ln\left\{\frac{[R_0]}{[R]}\right\} + ([R_0] - [R])$$

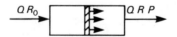

Fig. 4.8 Schematic representation of a plug flow reactor (PFR).

Kinetic properties and reactor design

Again, this can be expressed in terms of fractional conversion

$$\frac{V'_{max} \cdot V_{tot}\varepsilon}{Q} = K'_m \ln \left\{\frac{1}{1-X}\right\} + [R_0]X$$

$$\frac{V'_{max} \cdot V_{tot}\varepsilon}{Q} = [R_0]X - K'_m \ln\{1-X\} \qquad (17)$$

The performance equation for the PFR (equation 17) can be rearranged for the calculation of the kinetic constants from experimental data.

$$[R_0]X = K'_m \ln\{1-X\} + \frac{V'_{max} \cdot V_{tot}\varepsilon}{Q}$$

So a plot of $[R_0]X$ versus $\ln\{1-X\}$ will give a slope of K'_m and an intercept of $V'_{max} \cdot V_{tot}\varepsilon/Q$. As V_{tot}/Q is equivalent to $1/D$, this last term can be simplified to $V'_{max}\varepsilon/D$.

Choice of reactor type

One of the most frequently cited advantages of immobilization is that it allows the re-use of expensive enzyme catalyst. However, more significantly, the use of immobilized enzymes in continuous reactors allows a smaller reactor to be used.

In the food industry glucose isomerase is used in a continuous-flow packed-bed reactor to produce high fructose syrup. The residence time of liquid in the column is about 20 minutes. Taking this as a realistic residence time, the situation can be considered where 0.5 m³ of product solution was required in 10 hours. A comparison of the reactor size required for batch and continuous operation can be made:

A single vessel with a batch time of 10 hours would need to have a volume of 0·5 m³.

A continuous reactor having a residence time of 0.33 hours and a flow rate of 0.05 m³ hours⁻¹ would give a reactor volume of 0.0165 m³.

Obviously, these represent extreme cases, and the batch time could be reduced by using more enzyme. However, the shorter the batch cycle time, the more closely the system approaches continuous operation. The labour costs associated with repetitive batch operation would also need to be considered when assessing overall performance.

From these considerations it is apparent that continuous operation is desirable wherever possible. When running a continuous enzyme reactor a decision must be made between a well-mixed or plug flow mode of operation. The performance equations have been derived for both modes but their differences in terms of enzyme requirement, inhibition effects and implications of enzyme stability must also be considered (Vieth and Venkatsubramanian, 1973).

Enzyme requirement. Depending on the reactant concentration in the reactor, the kinetics shown could range from zero to first order. If these are considered as the extreme cases, then

Zero order $[R] \gg K_m'$

CSTR $\qquad \dfrac{V_{max}'}{D} = X[R_0] + K_m'\,[X/(1-X)]$

PFR $\qquad \dfrac{V_{max}'}{D} = X[R_0] - K_m'\,\ln\{1-X\}$

(N.B. Dilution rates for PFRs are now expressed in terms of the liquid volume, V_l)

When the reactant concentration is much higher than K_m', the reactors are operating in the zero-order region, and so performance in each case is approximated by

$$\dfrac{V_{max}'}{D} = X[R_0]$$

The enzyme requirement will, therefore, be similar for both PFR and CSTR.

First order $K_m' \ll [R]$

In this case the equations can be approximated by

CSTR $\qquad \dfrac{V_{max}'}{D} = K_m'\,[X/(1-X)]$

PFR $\qquad \dfrac{V_{max}'}{D} = -K_m'\,\ln\{1-X\}$

If the same immobilized enzyme preparation and the same dilution rate are used, then the ratio of the two equations gives

$$\dfrac{K_2 E_0 \text{cstr}/D}{k_2 E_0 \text{pfr}/D} = \dfrac{E_0 \text{cstr}}{E_0 \text{pfr}}$$

$$= \dfrac{K_m'\,[X/(1-X)]}{-K_m'\,\ln\{1-X\}}$$

$$= \dfrac{-X}{(1-X)\ln\{1-X\}}$$

This can be solved for any required fractional conversion to show the ratio of enzyme, required by the two reactors (Fig. 4.9).

Inhibition effects. The effects of inhibition can be determined by deriving the performance equations using the required kinetic expression. The common examples are given in Table 4.1. As the CSTR operates at outlet conditions, i.e. the reactant concentration in the reactor is the same as in the product stream, its performance is less affected by reactant inhibition than a PFR, where reactant

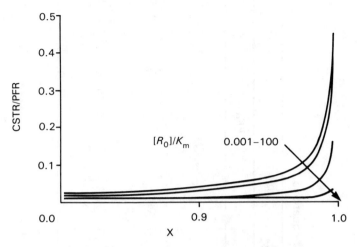

Fig. 4.9 Ratio of enzyme required in a CSTR compared with a PFR as a function of desired fractional conversion. Changes in $[R_0]/K_m$ indicates a change from first- to zero-order kinetics.

enters the reactor with no dilution. Conversely, product inhibition has a more pronounced effect on CSTRs as there will be an evenly distributed product level throughout the reactor. Even in PFRs where there is little product in the first part of the reactor, product inhibition has a significant effect on performance, in some cases increasing the enzyme requirement by several orders of magnitude.

Stability effects. Enzyme inactivation in a reactor is usually modelled using a first-order decay constant in a similar way to free enzyme (see Chapter 6). If the performance of a reactor at time zero and after some elapsed time t is compared, an estimate of the decay constant (k_d) can be obtained for CSTR.

$$\text{Time}^t \; \frac{V_{max}^{t'}}{D} = X^t \, [R_0] + K_m' \, [X^t/(1-X^t)]$$

$$\text{Time}^0 \; \frac{V_{max}^{0'}}{D} = X^0 \, [R_0] + K_m' \, [X^0/(1-X^0)]$$

Again, a ratio of the two equations can be simplified if k_2 and D are identical.

$$\frac{E_0^t}{E_0^0} = \frac{X^t \, [R_0] + K_m' \, [X^t/(1-X^t)]}{X^0 \, [R_0] + K_m' \, [X^0/(1-X^0)]}$$

The decay of free enzyme can be described by

$$\ln \frac{E_0^t}{E_0^0} = -k_d \cdot t$$

Table 4.1 Summary of common reactor performance equations

Type of inhibition	Kinetic expression	Reaction performance equation	
		CSTR	PFR
None	$v = \dfrac{V_{max} R}{K_m + R}$	$\dfrac{V'_{max}}{D} = R_0 X + K'_m \left[\dfrac{X}{1-X}\right]$	$\dfrac{V'_{max}}{D} = R_0 X - K'_m \ln[1-X]$
Reactant (uncompetitive)	$v = \dfrac{V_{max} R}{R\left[1 + \dfrac{R}{K_i}\right] + K_m}$	$\dfrac{V'_{max}}{D} = R_0 X + K'_m \left[\dfrac{X}{1-X}\right] + \dfrac{R_0^2}{K_i}(X - X^2)$	$\dfrac{V'_{max}}{D} = R_0 X - K'_m \ln[1-X] - \dfrac{R_0^2}{2K_i}(2X - X^2)$
Product (competitive)	$v = \dfrac{V_{max} R}{[R] + K_m\left[1 + \dfrac{P}{K_i}\right]}$	$\dfrac{V'_{max}}{D} = R_0 X + K'_m \left[\dfrac{X}{1-X}\right] + \dfrac{K'_m}{K_i} \dfrac{R_0 X^2}{(1-X)}$	$\dfrac{V'_{max}}{D} = R_0 X \left[1 - \dfrac{K'_m}{K_i}\right] - K'_m \ln[1-X]\left[1 + \dfrac{R_0}{K_i}\right]$

Kinetic properties and reactor design

Therefore

$$-k_d \cdot t = \ln \left\{ \frac{X^t [R_0] + K_m' [X^t/(1-X^t)]}{X^0 [R_0] + K_m' [X^0/(1-X^0)]} \right\}$$

Once a value for k_d has been determined, then the value of V_{max} can be predicted for any given time.

$$\frac{V_{max}^{t\prime}}{D} = \frac{V_{max}^{0\prime}}{D} \exp-(k_d t)$$

Similar equations can be derived for plug flow reactors to allow the effect of operation time on fractional conversion to be calculated once k_d is known.

In practice this approach must be used with caution as some enzyme inactivation is related to the volume of material passed through the reactor rather than a purely time-related function. Rather than make an exact prediction of performance, it is usual to monitor product concentrations and adjust the dilution rate to maintain the desired conversion. If the enzyme's stability is less than ideal, it may be necessary to consider the effects of stability when choosing the reactor type. A comparison of calculated k_d values shows that the performance of a PFR is more severely influenced by enzyme inactivation than a CSTR (Fig. 4.10). However, this must be set against the higher initial enzyme requirements of a CSTR for a given performance.

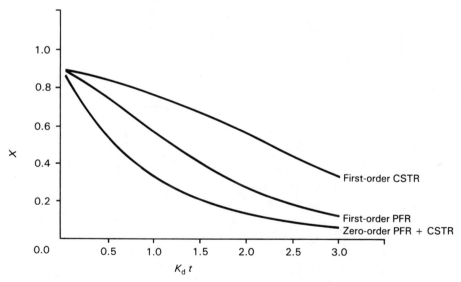

Fig. 4.10 Effects of inactivation on the performance of immobilized enzyme reactors.

Conclusions

This chapter has demonstrated that the choice of reactor operating conditions will be influenced by the kinetics and stability of the enzyme system to be utilized. Other factors which may need to be considered are the feed viscosity and the presence of particulate material in the feed which could lead to column plugging and an elevated pressure drop across the reactor bed. In situations where the fluid dynamics of the reactor are not known, a simple assessment of the operating mode can be made using tracer studies (see Appendix 2).

The sensitivity of enzymes to a wide range of environmental factors makes prediction of properties difficult. In contrast to the highly defined conditions used in pure biochemical studies, technological processes represent a highly complex situation where it is not possible to control all relevant parameters (Weetall and Pitcher, 1986). Care must be taken, therefore, to avoid placing too much emphasis on theoretical relationships and to ensure that experimental evaluations are carried out under appropriate conditions (Bucholz *et al.*, 1979). The examination of kinetic relationships made in this chapter has avoided a discussion of multi-reactant enzymes. The methods used to derive kinetic expressions in these cases are similar but obviously more complicated. As many enzymes of commercial interest are hydrolases, one reactant (water) is always in vast excess and the kinetic expression can be approximated to the simple Michaelis–Menten case.

Chapter 5

Medical and pharmaceutical applications of enzymes

Introduction

The variety of uses of enzymes in medicine and allied subjects is potentially immense although, at present, the number of applications is relatively small. Nonetheless, the results from this small group of successful ideas are exciting and demonstrate clearly the potential of the techniques involved. The medical and pharmaceutical application of enzymes covers such a diverse range of ideas that it is convenient to divide up this subject into three main areas of interest: enzyme therapy, analytical uses and the production of pharmaceutical compounds. Each area, although encompassing a wide spectrum of applications, has certain overriding principles that are central to the successful use of enzymes. In this chapter the main emphasis will be placed on the use of enzymes as therapeutic agents with correspondingly less space devoted to the two other topics.

In contrast to most industrial uses, medical and pharmaceutical applications generally require small quantities of relatively highly purified enzymes. In part, this reflects the fact that for an enzyme to be effective only the compound(s) of interest in a complex physiological fluid or tissue should be modified. This contrasts with most industrial processes where the feed-stock is relatively well defined and therefore a crude enzyme preparation may be used. Furthermore, if an enzyme or an enzyme-generated product is to be given to a patient, then it is clearly desirable that a minimum of extraneous material is administered in order to avoid possible side effects.

Enzyme Therapy

The concept of this form of therapy is simply the administration of a particular enzyme to a patient, hopefully resulting in an improved prognosis. The fundamental problem associated with this approach is that the defensive responses of the body inactivate or remove exogenous compounds. Consequently, any scheme involving the administration of an enzyme either via the blood (intravenously) or by any other route must take this problem into account.

The physiological response to the administration of a foreign material (antigen) is to produce specific antibodies against the compound. The ability to produce an antibody to a specific antigen is retained often for the lifetime of the individual; this is the basis of acquired immunity.

Another problem, particularly with the intravenous administration of an enzyme, is that proteins have a measurable half-life regardless of the immunological response of an individual. For instance, the half-life of an enzyme in the blood depends to a large degree on the glycoprotein nature of the molecule. Typically, glycoproteins with sialic acid-terminated oligosaccharide units will have relatively long half-lives whereas those terminated in mannose or galactose residues will be removed from the blood within minutes. Complete removal of the oligosaccharide units will give intermediate values for the half-life.

Clearly, one way to minimize the problems associated with therapy is to encapsulate the enzymes. This approach does have its problems; for example, the material used to encapsulate the enzyme may also be antigenic. Also, many synthetic materials can induce blood clotting, with disastrous results, unless anticoagulants are given simultaneously. However, with a careful choice of material for encapsulation, the problems can be minimized.

The strategy adopted for a therapeutic application depends on the type of problem to be solved. In general, chronic diseases require a different approach to acute conditions. By definition, the former is a long-term problem whereas the latter is short-term. Chronic conditions include the genetically inherited enzyme defects, organ failure and the treatment of certain tumours. Acute disorders include myocardial infarction (heart attack) and detoxification.

Genetic defects

More than one hundred and fifty metabolic diseases have been ascribed to specific enzyme defects. In many of the defects the enzyme is completely missing whereas in others it is replaced by a relatively inactive isoenzyme. Although a small proportion of the disorders, e.g. phenylketonuria (a defect in aromatic amino acid metabolism), can be treated by controlled diets (i.e. no phenylalanine), most of these inborn errors in metabolism result in physically and mentally crippling diseases. In theory, it should be possible to administer the 'missing' enzyme to the patient and to alleviate or remove the symptoms of the disorder (Beutler, 1981). However, in practice some of the damage, particularly to brain development, will be irreversible. Nevertheless, there is a significant group of these disorders that should be amenable to enzyme therapy. These are disorders in which an enzyme responsible for the degradation of a particular compound is missing, resulting in the accumulation of substrate within the lysosomes (the lysosomal storage diseases) (Tager, 1985).

Attempts to treat inherited metabolic disorders by the administration of exogenous enzyme have met with only partial success. The major problem is that the exogenous enzyme tends to be accumulated in the liver and spleen whereas other organs may still be deprived. An example of this problem is the treatment of Type II glycogen storage disease (Pompe's disease), a defect in glycogenolysis that is usually fatal within twelve months of birth. This disease is caused by the lack of an active α-1,4-glucosidase, resulting in the accumulation of large quantities of glycogen within the lysosomes of liver and muscle. Treatment of a seven-month-old child by intravenous injection of liposomes containing α-1,4-glucosidase resulted in some decrease in liver glycogen. However, this treatment did not reduce the quantity of glycogen stored in the muscles and the disease took its natural course (Hers and Barsy, 1973).

The observation that liposomes tend to accumulate preferentially in the liver lysosomes is the crux of the problem with enzyme therapy. Other methods of delivery also result in enzyme accumulation by the liver and spleen, with little or none reaching other organs. Considerable research effort is aimed at specifically targetting enzymes to ensure an appropriate distribution amongst the various organs and clearly there is some way to go before enzyme therapy can be considered to be an effective method of combating genetic defects.

Genetic disorders of a different type from those already described are much more amenable to enzyme therapy. Cystic fibrosis, which affects 1 in 1600 of Northern Europeans, often results in pancreatic insufficiency. In such cases the pancreatic ducts are blocked and the enzymes that are vital for normal digestion are unable to reach the intestine resulting in malnutrition. Oral administration of pancreatic enzymes, suitably coated to prevent denaturation by the stomach acid, helps to produce normal digestion and nutrition (Goodchild and Dodge, 1985).

Artificial organs

Artificial organs incorporating enzymes have been developed to replace some of the functions of the kidney and liver. Unless organ transplantation is available, chronic renal failure is treated by periodic haemodialysis; this process is time consuming (6–12 hours treatment three times a week) and requires the patient to be immobile. A potentially attractive solution is to reduce the amount of conventional haemodialysis and to use a small, portable, artificial kidney in the intervening periods (Chang, 1977).

Urea is the major toxic compound that accumulates in blood as a result of kidney failure. This compound can be degraded to carbon dioxide and ammonia by the enzyme urease and the products removed from the blood. Carbon dioxide will be removed naturally by expiration but the ammonia will have to be adsorbed onto a suitable material such as charcoal. Theoretical calculations, based on animal studies, show that an extracorporeal shunt (dimensions of 2 cm diameter × 10 cm long) containing nylon-microencapsulated urease and a charcoal-adsorbent would be sufficient to maintain the average adult (Fig. 5.1). Animal experiments have demonstrated that blood urea concentration is reduced by 50%

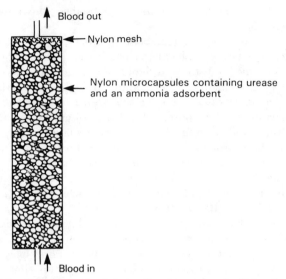

Fig. 5.1 Extracorporeal shunt for the removal of blood urea.

using these techniques, although the quantity of ammonia did increase. The present limitation to the system is the development of an adsorbent with sufficient capacity to bind all of the ammonia.

An interesting alternative to the extracorporeal shunt is to administer microencapsulated urease and adsorbent orally. Urea equilibrates across the intestinal wall and therefore depletion of the gastrointestinal-tract urea will result in a reduction in the concentration of blood urea. Experiments with animals have validated this approach and shown it to be almost as effective as the shunt without many of the side effects.

The liver is a multifunctional organ and it would be impossible to produce a substitute using present technology. However, one important function of the liver, detoxification, can be reproduced (Safer, 1979). Microsomal-bound enzymes from liver cells are able to detoxify a variety of compounds. Essentially, these enzymes catalyse hydroxylation, demethylation and conjugation reactions, thus rendering the toxic compound more water-soluble and therefore more excretable. Preparations of microsomal enzymes, e.g. cytochrome P450, have been shown to be effective in the detoxification of many compounds, e.g. barbiturates, although the requirement for NADPH and molecular oxygen may be a problem in artificial systems. Detoxification systems may be suitable either as a partial replacement of liver function or to act as an additional mechanism in cases of acute toxicity. Although the development of the artificial kidney and the detoxification systems is well under way, neither is yet ready for human use.

Control of neoplasms

The treatment of certain blood-related neoplasms with intravenously administered L-asparaginase has been successful particularly in the case of acute

Medical applications of enzymes

lymphocytic leukaemia (Cooney and Rosenbluth, 1975). Complete remission of the disease has been reported in up to 60% of the patients after enzyme treatment. Animal studies have indicated that this enzyme may be useful in the treatment of other forms of leukaemia and related disorders, although this has not yet been substantiated by clinical data.

The treatment of acute lymphocytic leukaemia with L-asparaginase is based on the observation that certain tumour cells have a nutritional requirement for exogenous L-asparagine whereas normal cells are able to synthesize this amino acid *in situ*. Therefore, intravenous administration of the enzyme will reduce blood levels of L-asparagine depriving the abnormal cells of a vital nutrient and cause a regression of the neoplasm. It has been suggested that the resultant relatively high levels of L-aspartate (a product of the enzyme reaction) may also be toxic to the tumour cells. A further advantage of this treatment is the high therapeutic index of the enzyme (1000) compared with alternative anti-leukaemic drugs (therapeutic index < 10). The therapeutic index is the ratio of toxicity (usually measured as the minimum lethal dose) to the normal therapeutic dose.

Originally, L-asparaginase was obtained from guinea-pig serum (very few mammals possess significant quantities of serum L-asparaginase). However, this source would not provide sufficient quantities for regular clinical use and L-asparaginase is obtained routinely from *Escherichia coli*. Preparations from other bacteria such as *Erwinia caratovora* and *Serratia marcescens* have also been used.

There are a number of important considerations when selecting a source of L-asparaginase. Naturally, it is essential that the enzyme will be active in the blood (i.e. pH 7.4, 37 °C). However, there are also some more subtle considerations; these are the catalytic efficiency of the enzyme and the rate of removal of this 'foreign protein' from the host's circulation. Blood concentration of L-asparaginase is low (4.3×10^{-5} mol l^{-1}) and therefore it is only enzymes with a low K_m value that are therapeutic (Table 5.1). To ensure adequate catalytic activity, enzymes with a low K_m are preferred as this minimizes the quantity of protein that has to be added to produce the desired effect (see Chapter 4 for an explanation of K_m and substrate concentration effects). It has been calculated that the serum concentration of L-asparagine must be reduced to less than 1×10^{-5} mol l^{-1} to cause regression of the neoplasm.

The rate of removal of the enzyme from the circulation is another important consideration. Preparations of L-asparaginase from various sources each have different serum half-lives (Table 5.1) and, in general, the longer the half-life, then the greater the efficacy of the enzyme. The degree to which the enzyme is removed from the circulation has been related to the isoelectric point of the preparation and is also dependent on glycoprotein structure. Successive injections of the enzyme also results in the generation of circulating anti-L-asparaginase antibodies that significantly decrease the serum half-life. The therapeutic use of L-asparaginase is not without side effects and these have included anorexia (absence of appetite), nausea, fever, diarrhoea and anaphylaxis (immunological hypersensitivity to the enzyme). One theory is that these side effects are a consequence of reduced concentrations of serum L-asparagine and L-glutamine (also a substrate for the enzyme), resulting in a reduced protein synthesis capacity even in normal cells.

Table 5.1 Properties of L-asparaginases for therapeutic use

Source	K_m value ($mol\ l^{-1}$)	Serum half-life[a] (hours)	Anti-neoplastic activity[b]
Escherichia coli			
EC I enzyme	n.d.[e]	Very rapid[d]	X
EC II enzyme[c]	1.2×10^{-5}	2–7	
Erwinia caratovora[c]	1.0×10^{-5}	4	
Serratia marcescens	1.2×10^{-5}	3–6	
Bacillus coagulans	4.7×10^{-3}	0.5	X
Guinea pig[c]	7.2×10^{-5}	26	
Agouti[c]	4.1×10^{-5}	11	

[a] Determined in mice. There is evidence, particularly with the EC II enzyme from *E. coli* that the serum half-life may be up to ten times longer in humans
[b] Determined in mice
[c] Used therapeutically in humans
[d] Exact value not specified
[e] Not determined

The side effects would also be typical of an immunological reaction against the enzyme.

The intravenous injection of L-asparaginase is a relatively crude, albeit successful, approach to therapy. Much effort has been concentrated on finding ways to reduce the immunological problems and to increase the serum half-life of the enzyme. Microencapsulation of the enzyme and extracorporeal shunts have proved useful in laboratory experiments but have not been used in a clinical context.

Blood circulation disorders
Various disorders in the blood circulation have been treated effectively with enzymes (Maciag et al., 1977). Urokinase, an enzyme isolated from human urine, has been applied successfully to the removal of blood clots. This enzyme causes the limited proteolysis of plasminogen to form plasmin which then dissolves the fibrin clot in the normal way. Urokinase is difficult to obtain in quantity and is therefore expensive to prepare because of difficulties in separation and concentration. However, these difficulties may be overcome by the application of techniques such as affinity chromatography and also by the cloning of the urokinase gene.

Streptokinase, a protein from haemolytic streptococci, also activates plasminogen. Despite its name, streptokinase is not an enzyme, although it is frequently assumed to be so. Streptokinase binds to plasminogen and this 1:1 complex is then able to activate other plasminogen molecules. Both urokinase and streptokinase have found considerable clinical use although the latter does produce side effects.

A novel enzyme treatment with tremendous potential is the treatment of myocardial infarction (heart attack) with purified preparations of hyaluronidase

(Hyalosidase or GL enzyme). A clinical trial in the UK has demonstrated that patients treated with a single intravenous dose of the enzyme within 6 hours of the onset of symptoms have significantly improved survival rates (Flint et al., 1982). Hyaluronidase degrades some of the polysaccharide components of connective tissues (glycosaminoglycans). It is argued that partial clearance of the glycosaminoglycans results in a reduction in oedema (water retention) and improved access of nutrients and oxygen to the damaged but recoverable tissues. As the treatment consists of a single dose, side effects are virtually nonexistent.

Attempts to provide a definitive explanation for the mechanism of action of hyaluronidase have produced some puzzling results. Animal studies have shown that more than 70% of the administered enzyme is accumulated in the liver within 10 minutes of injection and that only 0.1% of the total activity is associated with the heart. Biochemically, this would make sense as hyaluronidase is a glycoprotein containing three mannose-terminated oligosaccharides (the liver is the major site of uptake of this type of glycoprotein). Interestingly, it has been shown recently that damaged heart tissue can accumulate the enzyme preferentially and this may give a clue to the pharmacological activity of the preparation. Regardless of the possible mechanism(s) of action, there is no doubt that, with more than 100 000 deaths per annum from myocardial infarction in the UK, hyaluronidase is potentially an enormously important therapeutic agent.

Analytical uses

The analysis of physiological samples is central to modern medicine and enzymes (both free and immobilized) have a major part to play. Many physiological components are difficult to measure by conventional chemical assays largely because of a lack of sensitivity, selectivity or ease of operation. Also, some compounds, e.g. peptide hormones, are present in vanishingly small concentrations (10^{-12} mol l^{-1}), at which they have no directly measurable chemical or physical characteristics. The use of enzymes in analysis falls largely into two groups. The first of these is the direct enzymic assay of a compound whereas the second utilizes an enzyme to amplify another response, e.g. enzyme-linked immunoassays.

Direct analysis of metabolites

The specificity of enzymes makes them ideal tools for the analysis of a single compound in a complex physiological fluid. Numerous enzyme-based assays have been developed using a variety of detection methods (see Chapter 8). Two of the most significant enzyme assays are for the analysis of D-glucose and cholesterol in physiological fluids. Both compounds are difficult to measure with absolute specificity using conventional chemical methods.

D-Glucose may be quantified using glucose oxidase (obtained from *Aspergillus niger*) and horseradish peroxidase. The glucose oxidase reaction utilizes molecular oxygen to oxidize the D-glucose with the production of hydrogen peroxide (equation 1).

$$\text{D-glucose} + H_2O + O_2 \xrightarrow{\text{glucose oxidase}} H_2O_2 + \text{D-glucuronolactone} \quad (1)$$

The liberated hydrogen peroxide is assayed using the peroxidase enzyme and a suitable redox dye and the coloured product is measured spectrophotometrically (equation 2).

$$H_2O_2 + \text{reduced dye} \xrightarrow{\text{peroxidase}} \text{oxidized dye} + H_2O + \tfrac{1}{2}O_2 \quad (2)$$
$$\text{(colourless)} \qquad\qquad \text{(coloured)}$$

A wide range of suitable redox dyes is available for this assay. Glucose oxidase is effectively totally specific for D-glucose in physiological fluids. Although the enzyme has some activity towards D-mannose (20%) and 2-deoxy D-glucose (20%), neither of these compounds is a significant contaminant in the analysed samples.

The measurement of free cholesterol is analogous to the method decribed above for D-glucose. Cholesterol oxidase (obtained from *Nocardia* sp.) is used to oxidize cholesterol and produce hydrogen peroxide (equation 3) which can then be assayed as described previously.

$$\text{cholesterol} + H_2O + O_2 \xrightarrow{\text{cholesterol oxidase}} \text{cholest-4-en-3-one} + H_2O_2 \quad (3)$$

The sample may be pre-treated with cholesterol esterase (equation 4) if it is desired to measure total free and bound cholesterol.

$$\text{cholesterol ester} \xrightarrow{\text{cholesterol esterase}} \text{cholesterol} + \text{fatty acid} \quad (4)$$

These enzyme-based assays are available commercially in 'kit-form' and also they have been adapted to run on autoanalyser systems.

Indirect enzyme-linked assays
The measurement of many compounds of medical or pharmaceutical interest has been revolutionized by the introduction of immunological techniques. The remarkable specificity of antigen–antibody reactions lends itself to the development of sensitive and accurate analytical methods. To produce a suitable assay it is necessary to modify the antibody molecule so that it becomes possible to quantify the binding to the antigen. In the past, antibodies have been labelled with ^{125}iodine (β- and γ-ray emitter) and the bound radioactivity measured. This technique (radioimmunoassay), although tremendously successful, requires both expensive measuring equipment and strict safety precautions.

A recent advance has been to link covalently simple, hydrolytic enzymes to the antibody and to use the catalysed reaction to quantify the system. As the enzyme is a catalyst, it effectively amplifies the effect and it is possible to develop simple, safe and sensitive assays. For various technical reasons, it is advantageous to link the enzyme to a species-specific second antibody and to use a multilayer assay system (Fig. 5.2). Conventionally, the enzyme-linked immunosorbent assay system (ELISA) uses enzymes that have chromogenic substrates (peroxidase, alkaline phosphatase or β-galactosidase) and the progress of the reaction is monitored

Medical applications of enzymes

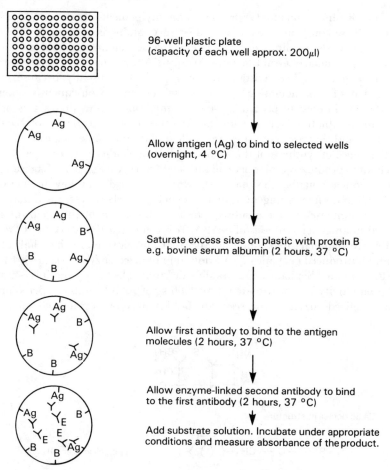

Fig. 5.2 Outline of a typical ELISA procedure.

spectrophotometrically. A newer idea is to use urease coupled to the antibody; in this system the hydrolysis of urea causes a pH change which can be monitored as an end-point by using a suitable indicator.

Although ELISA is a relatively new analytical method it has been so successful that virtually every biochemical or medically-based laboratory relies on the technique. The versatility of the system is virtually limitless given the number of possible antibody–antigen interactions that exist.

Pharmaceutical applications

Semi-synthetic antibiotics
The major pharmaceutical products produced using enzyme technology are the semi-synthetic penicillins. Conventional fermentation is used to produce either

benzyl penicillin (penicillin G) or phenoxymethylpenicillin (penicillin V) and in the past these compounds have been used successfully as antibiotics. However, both compounds are limited in their effectiveness against certain pathogenic bacteria (particularly Gram negative strains). Also, numerous bacterial strains have emerged that are resistant to these penicillins, usually because of the production of a β-lactamase (penicillinase) enzyme. Consequently, there is considerable interest in producing semi-synthetic penicillins that act as broad-spectrum antibiotics (suitable for the treatment of many types of bacterial infection) and that are more resistant to degradation by β-lactamases.

The route of synthesis of the semi-synthetic penicillins is to remove the aryl sidechain to produce 6-aminopenicillanic acid (6-APA) (Fig. 5.3) (Lagerlöf et al., 1976). Conventionally, this has been done chemically at low temperatures ($-40\,°C$) and under stringent anhydrous conditions. However, the discovery of penicillin amidases and the subsequent immobilization of these enzymes has resulted in an easier synthesis of 6-APA. It is estimated that $8-10 \times 10^3$ tons of 6-APA are manufactured annually and that currently more than half of this amount is produced enzymically. Various immobilized enzyme systems are in use (CNBr–activated Sephadex, glutaraldehyde cross-linked cells) and it is estimated that productivity is approximately 100–2000 kg of 6-APA produced per kilogram of immobilized enzyme. The resultant 6-APA may be acylated chemically to

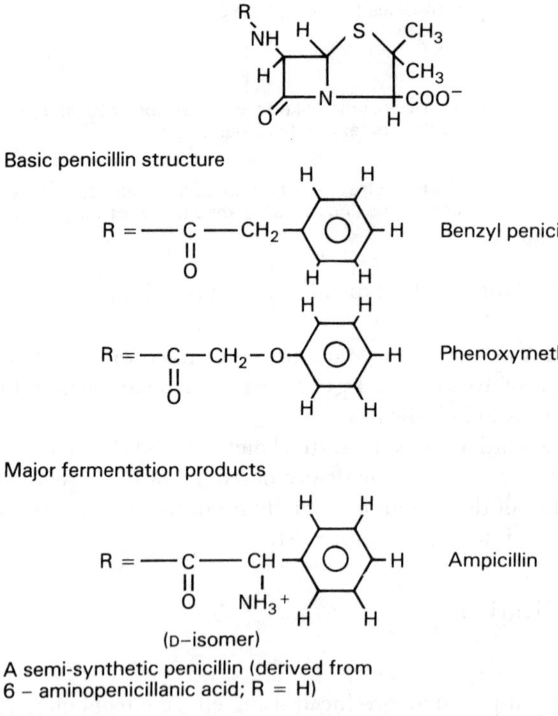

Basic penicillin structure

R = —C—CH$_2$—〈benzene〉—H   Benzyl penicillin (Pen G)
      ‖
      O

R = —C—CH$_2$—O—〈benzene〉—H   Phenoxymethyl penicillin (Pen V)
      ‖
      O

Major fermentation products

R = —C—CH—〈benzene〉—H   Ampicillin
      ‖ |
      O   NH$_3^+$
         (D-isomer)

A semi-synthetic penicillin (derived from 6 – aminopenicillanic acid; R = H)

Fig. 5.3 Production of semi-synthetic penicillins.

Medical applications of enzymes

produce a large range of different derivatives, e.g. ampicillin (Fig. 5.3), which have proved to be particularly efficacious.

The cephalosporins are a class of antibiotics with a structure similar to that of the penicillins (Fig. 5.4). The primary fermentation product, cephalosporin C, is not particularly useful as an antibiotic but the semi-synthetic derivatives have been immensely successful. Cephalosporin C is converted enzymically into 7-amino cephalosporonic acid which may then be acylated chemically to produce the cephalosporin-group of antibiotics, e.g. cephaloridine. The ability to alter the substituents at two different positions on the parent molecule greatly increases the possible range of compounds (Fig. 5.4).

Steroids
Steroids are used in a wide range of pharmaceutical preparations (e.g. the contraceptive pill, and anti-inflammatory agents) and processes for the production of these materials are of considerable economic importance. The chemical synthesis of steroids is fraught with difficulties because of the complex stereo-

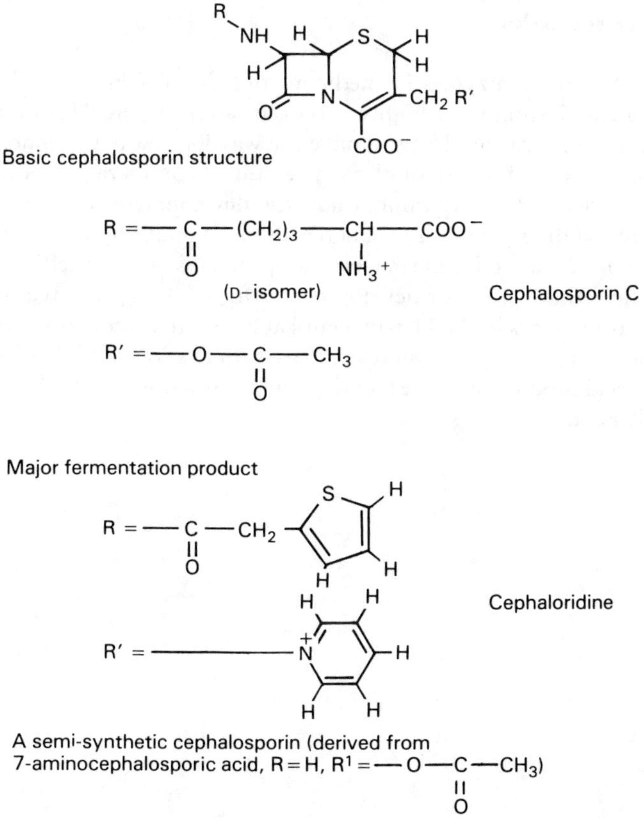

Fig. 5.4 Production of semi-synthetic cephalosporins.

chemistry of these compounds; many steps are required and overall yields are pathetically small.

More recently, microbial fermentations of diosgenin and stigmasterol have improved dramatically the yields of pharmaceutically important steroids compared with the chemical methods. However, diosgenin, in particular, is available in limited quantities and there is a real risk of supplies becoming exhausted. In contrast, other steroid precursors are readily available as byproducts of unrelated processes, although there is no method to ferment these to pharmaceutically useful end products. The potential of specific enzyme reactions promises to facilitate the utilization of the various precursor steroids that are readily available. Laboratory and pilot-plant scale experiments have demonstrated the feasibility of the approach, although no commercial process exists at present. One of the more significant problems is that many steroids are insoluble in aqueous solutions and therefore organic phases are essential. Nevertheless, enzyme reactions can work in non-aqueous sytems (see Chapter 10) and there is no reason why a satisfactory process cannot be developed.

Concluding remarks

At present the use of enzymes in medicine and the pharmaceutical industry is limited to a small number of highly successful applications. However, the very success of these applications helps to pave the way for new developments and it is clear that there is no shortage of ideas. The ethical considerations, quite rightly, make it more difficult to experiment and the development of suitable products takes a correspondingly long time. There can be no doubt that the application of enzymes to medical and pharmaceutical problems is an exciting and most promising field that is ripe for development. One of the great advantages of the application of enzymes in the pharmaceutical field is that economic processes can be developed on a relatively small scale. This is because of the high added value of pharmaceutical products and the fact that chemical syntheses are often multistage and complicated.

Chapter 6

Effects of immobilization on enzyme stability and use

Introduction

In any assessment of the pros and cons of a biological system in a technological context the problems of limited stability will come high on the debit side. It is clear that stability will be inextricably linked with process costs and as such is a central factor which must be considered in any assessment of feasibility. For a 'pure' biochemist stability is perhaps an artifical concept; whereas it is significant in the context of experimental procedures. It is difficult, if not impossible, to relate *in vivo* stability to that observed *in vitro*. However, the biotechnologist is more concerned with methods of utilization than conditions having biological relevance and so is able to investigate a wide range of approaches to influence the stability or apparent stability of the enzyme of interest.

Enzyme stability

The stability of a given enzyme will be a complex function of the environmental conditions used. Stability will vary with pH, reactant concentration and the presence of destabilizing agents. Generally the decay of enzyme activity is attributed to thermal effects, with the rate of decay being first order and reflecting the properties of the enzyme and its local environment (Cornish–Bowden, 1979).

$$\frac{d[E]}{dt} = -k_d [E_0]$$

This equation can be integrated to give the active enzyme concentration at time t ($[E_0^t]$):

$$[E_0^t] = [E_0^0] \exp(-k_d.t)$$

So the decay constant (k_d) for the enzyme can be determined from a graph of $\ln([E^t]/[E_o])$ versus t.

For an enzyme to be suitable for a commercial application, its stability must be sufficient for the purpose. In the case of a reactor system the stability will be quantified in terms of the profit made on the product formed during the lifetime of the enzyme catalyst. For a sensor the criterion would usually be based on the need for a linear response over an extended time period. This period would be determined by the cost of replacement and the run time of the process to be monitored. In addition to operational stability, ease and cost of storage must be considered.

The decay constant k_d will be related to temperature according to the Arrhenius relationship, with the activation energy of the destabilization process being critical. The presence of stabilizing compounds and the storage pH may change the activation energy and hence the necessary storage temperature.

The assessment of stability, both for storage and operation, will usually be in terms of half-life, i.e. the time taken for half of the enzyme activity to be lost.

$$[E^t] = \frac{[E_o]}{2} \quad \text{therefore} \quad \frac{[E_o]}{2} = [E_o] \exp(-k_d.t)$$

therefore

$$\ln(0.5) = -k_d t \quad \text{and} \quad t_{\frac{1}{2}} = \frac{0.693}{k_d}$$

To assess mechanisms of enhancing enzyme stability the conformational structure of the protein must be considered. This is the three-dimensional spatial arrangement adopted by the protein molecule and results from interactions between the sidechains of amino acids removed from each other in the primary sequence. Any compounds whose presence affects these interactions will have an effect on stability.

The major forms of interactions include hydrogen bonding, hydrophobic interactions and ionic bonds. Factors affecting these interactions include organic solvents, extremes of pH, detergents, and high salt concentrations (e.g. ammonium sulphate).

In addition to the above mechanisms, the formation of disulphide bridges between two cysteine residues is important in many extracellular enzymes. This covalent binding can be critical for enzyme activity and protecting this bridge from reduction can significantly increase the half-life of some enzymes. Conversely, many intracellular enzymes require the presence of reduced sulphydryl groups; in this case protection from oxidation is important.

A summation of the forces of conformation interactions may well give a value in excess of 400 kJ mol^{-1} but the degree of compensation is such that the free energy of activation for a globular protein is rarely greater than 60 kJ mol^{-1}. Thus native proteins are usually only marginally stable. When considering the wide range of environmental factors affecting an enzyme used in a technological context, it is apparent that theoretical prediction of 'operational stability' would be impossible, and that accumulation of experimental data is the only useful method of evaluation. Storage conditions used prior to the transfer of enzyme to the operational system can be much more rigidly controlled and represent a much simpler system to study (Buchholz, 1982).

Stabilization for storage
A distinction must be made between stabilization for storage and stabilization for operation. In the case of a high throughput enzyme reactor, cost factors might preclude the continuous addition of a stabilizer which may well be suitable for use during storage.

Most enzymes have been shown to be moderately stable in the range 0–4 °C. During storage when catalytic activity is not important this would represent the ideal temperature range. In some cases the presence of stabilizers, e.g. glycols and sulphydryl compounds, have been shown to be highly beneficial. The presence of reactants or reactant analogues has also been shown to have a stabilizing effect on many enzymes. This is thought to result from conformational changes leading to a more rigid structure after binding. Other agents shown to have stabilizing effects include organic polymers, anti-oxidants and chelating agents (Wiseman, 1978).

Stabilization for use
Methods of approaching the problem of operational stability can be divided into five categories:

(1) screening for naturally stable enzymes
(2) addition of stabilizers
(3) chemical modification
(4) immobilization
(5) protein engineering

Intrinsically stable enzymes are generally found in organisms adapted to life in hostile environments. There is now great interest in the enzymes from organisms adapted to life in hot springs (thermophiles) and highly saline environments (halophiles), and there has been a great deal of study on the structure–stability relationship of enzymes from thermophilic organisms. Although no clear principle has emerged, it appears that a degree of flexibility, allowing rapid renaturation, may be important. The greater stability of these enzymes must be balanced against the often higher cost of production when they are assessed for commercial applications.

The use of stabilizing additives must be carefully considered with respect to operational costs. However, in many cases reactant concentration can significantly influence enzyme stability and so the operational reactant concentration chosen may well restrict the choice of conditions and reactor configuration. The use of low concentrations of chelating agents, e.g. EDTA, microbial growth inhibitors (azide) and thiol protectors (hydrogen sulphide) may be beneficial, although in many cases these additives would be unacceptable on health and safety grounds.

Chemical modification of proteins includes acylation, alkylation, reactions with amino acid derivatives and other miscellaneous substituents. This is a complicated area where it is difficult to predict effects on an *a priori* basis. Effects on activity, stability and specificity have been demonstrated with various preparations and are considered in greater detail in Chapter 9. Again, the cost of modification must be balanced against the advantages.

Immobilized enzymes may show greater stability as a result of coupling to an insoluble support, although this cannot be assumed. In practice it is possible to improve stability by co-immobilizing specific chemical residues (e.g. sulphydryl groups, albumin) with the enzyme. Protection from the action of proteolytic enzymes and microbial degradation may be achieved by the use of entrapment methods.

Protein engineering is the most recent approach to improving enzyme stability. The methods used stem from advances in techniques of genetic engineering and computer graphics coupled with a greater understanding of protein structure (see Chapter 9).

Immobilization of enzymes

In considering the properties of enzymes it is also important to look at the implications of any modifications made to the enzyme structure. From a process point of view it may be desirable to 'immobilize' the enzyme to allow its retention in a reactor (Goldstein and Katchalski–Katzir, 1979). This immobilization is bound to lead to changes in the environment of the enzyme, and hence cause changes in the observed properties. The type and magnitude of these changes will depend on the enzyme and the method of immobilization used.

By 1972 the proliferation in the methodology of immobilization led to an attempt to rationalize the classification of immobilized enzyme systems. The major groups described are depicted in Fig. 6.1. Techniques for enzyme immobilization are now well established and there is a large body of literature describing this work. A brief description of the main types of immobilization is provided, but for a more detailed analysis the reader is referred to Trevan (1980).

Enzymes can be adsorbed onto a support surface. This process stems from physical rather than chemical interactions (e.g. charge effects and hydrophobic interactions) and the flexibility of this type of immobilization minimizes the

Adsorption

Covalent binding

Microencapsulation

Entrapment

Crosslinking

Fig. 6.1 Common methods of enzyme immobilization.

Effects of immobilization on enzyme stability

chances of structural distortion of the enzyme. Adsorption is regarded as being analogous to membrane-bound enzymes *in vivo*. The nature of the interactions is such that adsorption is usually a reversible process and changes in process conditions, e.g. pH and ionic strength, can cause desorption. Despite this disadvantage adsorption has been used for some industrial processes (e.g. glucose isomerase on DEAE–cellulose has been used commercially).

Covalent attachment of soluble enzymes to an insoluble support is the most common method for the immobilization of enzymes, and a wide range of techniques and supports have been used for this purpose. Although some activity may be lost during the coupling process, enzyme leaching from the support is unlikely with covalently bound preparations, ensuring a more stable catalyst material.

As an alternative to attaching the enzyme to a support, a method of entrapment can be adopted. In this case the enzyme remains in an unmodified state, but is trapped in a droplet of solvent within a polymer matrix. In the case of entrapment the enzyme will be dispersed throughout a polymer preparation, e.g. polyacrylamide. For encapsulation a more controlled approach is undertaken and the enzyme is entrapped in small capsules. One method which has been used is interfacial polymerization of a nylon membrane around an enzyme-containing droplet. The disadvantages of these approaches are the effects of the polymer on reactant availability. This prevents the use of entrapped enzymes with high molecular weight reactants. However, the polymer does give the enzyme protection under some circumstances. In particular, microencapsulated enzymes have been used *in vivo* where the capsule protects the enzyme from the host's immune system (see Chapter 5).

Although other methods of immobilization have been studied (e.g. ultrafiltration membranes), adsorption and covalent binding represent the major methodologies. More detailed reviews of methods of immobilization have been presented by other authors (Messing, 1985; Woodward, 1985).

Effects of immobilization on enzyme activity
When an enzyme is immobilized the activity of the preparation is invariably different from the native form. The reasons for this are many but it is possible to identify three major influences, namely, conformational, partitioning and diffusional effects (Goldstein, 1976).

Immobilization may cause structural perturbations in the protein, reducing the catalytic efficiency of an enzyme. In addition, immobilization may restrict the access of reactant to the enzyme's active site, again reducing activity. Both of these phenomena may be described as conformational effects.

Partitioning effects occur when the concentrations of reactant or other activity-related compounds may be different at the surface of the support matrix to those observed in the bulk solution. These effects can arise from electrostatic or hydrophobic interactions and, depending on the component(s) involved, partitioning effects may enhance or depress the observed rate of reaction.

Diffusional or mass transfer effects can be of two types, internal or external (Horvath and Engasser, 1974). External resistance comes from the presence of an unstirred liquid film surrounding the immobilized enzyme particle. If the rate of catalytic removal exceeds the rate of diffusion of reactant into this layer, then the reactant concentration at the surface will be lower than in the bulk solution. The thickness of this unstirred layer can be correlated with the flow properties of the solution passing over the particle. Internal mass transfer resistance arises when enzymes are entrapped in polymers. Again, the rate of reaction can remove reactant faster than it diffuses into the gel. This observed rate is a function of the diffusivity of the reactant, porosity of the polymer, external reactant concentration and the kinetics of the enzyme.

Conformational changes are the most difficult to assess and have not been as well described as partitioning and mass transfer effects. Also, conformational effects will vary with the enzyme, coupling method, and the support that is used. Some of the adverse consequences of immobilization on enzyme activity are illustrated in Fig. 6.2.

One of the most significant implications of partitioning effects is the potential to control the pH in the vicinity of the enzyme independently of the pH of the bulk solution. If an enzyme is immobilized on a charged support, e.g. an ion-exchange resin, the difference in the concentrations of charged components between the surface and the bulk phase can be described by a partition coefficient (P).

$$P = \frac{C_s}{C_b} \tag{1}$$

where C_s = concentration at the surface
C_b = concentration in the bulk solution.

In the case of H^+ ions, these effects have been correlated with a Boltzmann-type distribution.

$$H_s^+ = H_b^+ \exp(-e\psi/kT) \tag{2}$$

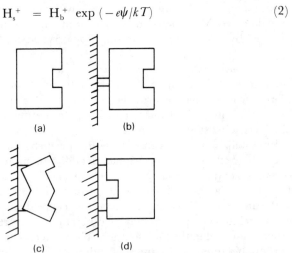

Fig. 6.2 Consequences of enzyme immobilization: (a) free enzyme; (b) active immobilized enzyme; (c) inactive immobilized enzyme (conformational change); (d) inactive immobilized enzyme (sterically hindered).

where H_s^+ = hydrogen ion concentration at the surface
H_b^+ = hydrogen ion concentration in the bulk solution
e = electronic charge
ψ = electronic potential
k = Boltzmann constant
T = absolute temperature

This can be expressed in terms of a partition coefficient by dividing though by H_b^+.

$$P_{H^+} = \exp(-e\psi/kT) \qquad (3)$$

These equations can be used to predict the surface hydrogen ion concentration if the properties of the solution and the support are known. For a polyanionic support, the surface hydrogen ion concentration would be higher than that measured in the bulk solution whereas for a polycationic support it would be lower.

The pH change can be calculated from equation (2) by dividing through by H_b^+ and taking the logarithm of both sides.

$$\ln\left\{\frac{H_s^+}{H_b^+}\right\} = (-e\psi/kT)$$

$$2.303 \log\left\{\frac{H_s^+}{H_b^+}\right\} = (-e\psi/kT)$$

$$pH = 0.43\,(e\psi/kT)$$

As the pH activity profile of an enzyme results from the dissociation of critical groups at the active site, this relationship can be used to calculate the change in the observed pH optima of an immobilized enzyme. As each enzyme will tend to have its own distinct pH optimum, processes based on a sequence of enzymes create problems in choosing an appropriate pH. However, by modifying the charge properties of the supports used, each enzyme can be immobilized at its own pH optimum, ensuring maximal activity. A similar approach can be taken with ionizable reactants to calculate the charge effects on surface reactant concentration. It should be noted that at high ionic strength the pH microenvironment effect is abolished (Fig. 6.3).

The effects of external mass transfer resistance can be modelled by assuming a linear gradient of reactant concentration across the unstirred boundary layer. This leads to a modified K_m and V_{max} term.

$$\frac{-d\,[R]}{dt} = \frac{V'_{max}\,[R]}{[R] + K'_m}$$

where V'_{max} = observed V_{max}
$[R]$ = bulk reactant concentration

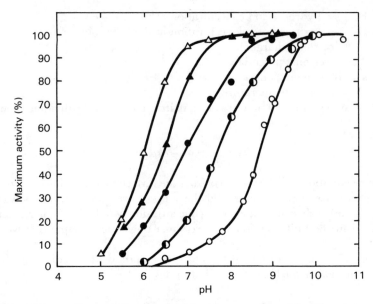

Fig. 6.3 Effects of immobilization and ionic strength on the pH activity curves for trypsin and a polyanionic, ethylene-maleic acid (EMA) copolymer derivative of trypsin (EMA-trypsin) at different ionic strengths: Trypsin—△, 3.5×10^{-2}; ▲, 1.0; EMA-trypsin—○, 6.0×10^{-3}; ◐, 3.5×10^{-3}; ●, 1.0. The substrate was benzoyl-L-arginine ethyl ester.
(From Goldstein, 1976)

$$K'_m = \frac{V'_{max} \delta}{D_s}$$

D_s = diffusivity of the reactant
δ = boundary layer thickness

The boundary layer thickness can be expressed in terms of the superficial liquid velocity in packed beds and the agitation rate in stirred tanks. As the thickness of the boundary layer is a conceptual aid to modelling and lacks physical significance it is usually expressed in terms of a mass transfer resistance (K_s).

$$K_s = \frac{D_s}{\delta}$$

This can be calculated from the Sherwood number (Sh) which is a dimensionless group relating total mass transfer in the system to the mass transfer arising solely from molecular forces (diffusivity).

$$Sh = \frac{K_s d_p}{D_s}$$

where d_p = particle diameter.

Effects of immobilization on enzyme stability

The Sherwood number can also be calculated using correlations with the fluid dynamics of the system.

$$Sh = cRe^a Sc^b$$

where a, b and c = constants whose value depends on the value of the Reynolds number (Re)
Sc = the Schmidt number

For values of the Reynolds number $(20 < Re < 120)$, the following constants may be used, a = b = 0.33 and c = 4.6.

The Reynolds number can be interpreted as the ratio of momentum or inertial forces to the viscous forces in the system. Hence a low Re number (< 100) indicates a viscosity-limited system which will exhibit laminar flow, whereas a high Re number (> 2000) represents fully turbulent flow (McCabe et al., 1985).

Reynolds numbers are defined according to the reactor type. Thus, for fixed beds

$$Re = \frac{\mu d_p \rho}{v}$$

and, for stirred vessels,

$$Re_i = \frac{(n d_i) d_i \rho}{v}$$

where d_i = impeller diameter
μ = liquid velocity
d_p = particle diameter
ρ = density of liquid
v = dynamic viscosity
n = stirrer speed
d_i = diameter of the impeller

The Schmidt number is the ratio of diffusivity over viscosity.

$$Sc = \frac{D_s \rho}{v}$$

So the Sherwood number can be calculated for the experimental conditions used and can be solved for the mass transfer rate. This can be used with the measured V_{max} to quantify the kinetic expression.

The effect of internal mass transfer on the observed rate of reaction again results from a reduction in the available reactant concentration. At high enzyme loadings the rate of consumption of reactant is faster than the rate of reactant diffusion into the polymer, leading to a concentration gradient within the immobilization matrix. This phenomenon can be described mathematically in terms of a mass balance over an infinitely small element of the polymer. The complexity of the resultant expression is such that a simple analytical solution is possible only for systems showing first-order kinetics. In practice, the effects of internal mass transfer are usually calculated using generalized effectiveness factor plots, where the ratio of V_{max} (observed)/V_{max} (theoretical) is plotted against $K_m/[R]_o$, where $[R]_o$ is the concentration of reactant entering the reactor, for a range of dimensionless Thiele moduli (ϕ). The Thiele modulus is of the form

$$\phi = L\sqrt{\frac{V_{max}}{K_m D_e}}$$

where L = half the thickness of the particle for the external mass transfer limited immobilized enzyme

D_e = diffusivity of the reactant through the immobilization matrix

In the case of internal mass transfer resistance there is no direct effect of external fluid dynamics. (Indirectly the effect of external resistance must be considered in series to internal resistance.) Internal resistance results from the small size and convoluted pathway of pores in the polymer support preventing a forced flow of solution through the beads. This results in a reduction of the observed diffusivity from the value obtained in free solution (Furui and Yamashita, 1985; Hannoun and Stephanopoulos, 1986). The effective diffusivity in the polymer phase can be expressed as

$$D_e = \frac{D_s \chi}{\tau}$$

where χ = the porosity

τ = tortuosity (path length)

As might be expected, the effective diffusivity can be shown to be proportional to the water content of the support and inversely proportional to the size of the reactant. The effectiveness factor plot (Fig. 6.4) is constructed by numerically integrating the mass balance for reactant across the particle (Horvath and Engasser, 1974). It can be seen from the plot that, when ϕ is small, the effectiveness factor tends to unity, but that ϕ increases with increasing particle size. From the $K_m/[R]$ lines it is apparent that a high reactant concentration will tend to over-ride the effects of internal mass transfer resistance. However, the use of elevated reactant concentrations may lead to a lowering of fractional conversion. In a packed-bed reactor the reactant concentration will vary along the reactor length and so the effectiveness factor will be a function of axial position in the reactor.

The implications of internal mass transfer on enzyme loading suggest that a compromise between a highly active preparation showing a low effectiveness factor and a low activity preparation showing a high effectiveness factor must be made.

Although internal mass transfer is generally undesirable, as it leads to a lowering of effective reactant concentration, it can be seen that, in systems showing reactant inhibition, this may offer some advantage (Atkinson, 1985). The collective effects of enzyme immobilization and activity are illustrated in Fig. 6.5.

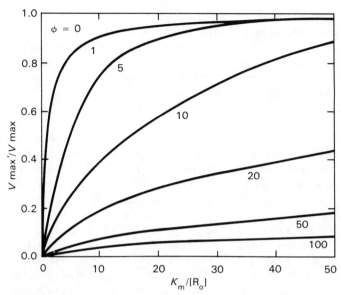

Fig. 6.4 Thiele modulus plot for internal mass transfer resistance. Plots of the overall rate of reaction in an enzymic membrane, V'_{max}, normalized to V_{max}, against the dimensionless concentration of the substrate at the surface, $K_m/[R_0]$, at different values of the Thiele modulus, ϕ. The effect of diffusion limitations increases with increasing values of ϕ, and results in a decrease of the overall rate of reaction with respect to the kinetically controlled rate which is obtained at $\phi = 0$.
(From Horvath and Engasser, 1974)

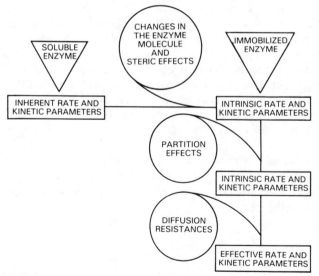

Fig. 6.5 Schematic illustration of factors affecting the activity of an immobilized enzyme preparation.
(From Engasser and Horvath, 1976)

Effects of immobilization on stability

Increased stability is often quoted as an advantage of enzyme immobilization. From a superficial examination of the literature concerning immobilized enzymes, it is easy to conclude that immobilization almost invariably leads to stabilization. However, it has been suggested that examples of stabilization are preferentially reported and can lead to a biased picture of the true situation. As with free enzymes, a distinction can be made between storage and operation stability. Operational stability is obviously the most important criteria for an immobilized enzyme. Although measurements of storage stability are far easier to obtain than operational data, it is clear that there will be no absolute correlation between the two situations and that storage stability data are of little value in predicting the operational stability of a given system (Vieth and Venkatasubramanian, 1973).

Although it is impossible to derive a mathematical expression for the effect of immobilization on stability, certain trends can be seen. In the case of proteolytic enzymes, a stabilizing effect is often observed and can usually be attributed to a reduction in the degree of autolysis. Unlike soluble enzymes, which usually show a first-order decay process, the decline in activity of immobilized enzymes often shows a polyphasic rate, i.e. an initial slow phase of decline followed by a period of rapid decay (Cheetham, 1983). One explanation of this phenomenon is that immobilization techniques often lead to a heterogeneous population of enzyme molecules in different orientations. Depending on exact points of immobilization, the stability of each population may vary. This is supported by evidence that enzymes which are entrapped rather than coupled show similar stability to the native enzyme. The complexity and number of coupling permutations possible with immobilization limits the potential for predictive relationships and it is generally regarded that experimental data must be obtained under the process conditions envisaged. In addition to the effects of immobilization on enzyme structure, the mass transfer limitations discussed earlier, if not accounted for, can lead to anomalous stability data. For example, a sample with a high initial activity but low effectiveness factor will show an initially slow decrease in performance with time. However, once a critical point is passed, the fall-off of performance becomes very rapid. This gives a biphasic decay of performance although the rate of enzyme decay would be constant. Similarly, anomalous increases in performance for diffusion-limited systems may be obtained if attrition of the immobilized enzyme leads to a decrease in particle size, and hence decreased mass transfer resistance.

The theory described for partitioning effects of pH activity profiles also applies to stability. If a process must be run at a non-ideal pH, the enzyme's stability could be improved by using an appropriately charged support giving the optimal local environment for the enzyme.

Conclusion

In many cases immobilization appears to confer some stability advantages on an enzyme. In practice, these are impossible to predict on an *a priori* basis. Under some conditions apparent increases in stability may be artefacts which result from a masking of the true situation by mass transfer effects. The overall complexity of the factors affecting enzyme stability suggest that, at present, the only feasible approach is to obtain experimental data under the conditions to be used, and a recent monograph on the characterization of immobilized biocatalysts proposes a standardization of the experimental approach to such stability assessments (Goldstein and Katchalski-Katzir, 1979).

Chapter 7

Uses of enzymes in agriculture and the food industry

Introduction

The earliest example of the use of enzymes probably results from the fortuitous observation that milk stored in the stomach sacs of goats was precipitated and that the resultant curds were a useful food material. As with many other areas of biotechnology, the early use of this reaction did not require any profound understanding of its mechanism and it was not until relatively recently that a more structured approach was taken to the use of enzymes. Other natural processes based on enzyme activity include the tenderization and flavour enhancement of meat during storage after slaughtering, the flavour changes associated with cheese ripening, and the release of fermentable sugars from barley during malting. The large expansion in the application of enzymes in the manufacturing and processing industries has largely been in two areas: the enhancement of traditional processes, and the development of totally new uses based on an understanding of enzyme properties. In this chapter applications of enzymes in agriculture and the food industries will be considered, together with the economic implications of these processes.

Enhancement of traditional processes

Milk products
The major milk-derived food product is cheese, which is produced by precipitating and recovering the casein component of milk. This precipitation can be achieved in two ways. First, the pH of the milk can be lowered to the isoelectric point of casein (4.6) by encouraging bacterial conversion of lactose to lactic acid. Second, the casein can be split enzymically to induce coagulation (Taylor et al., 1979). In practice, the method adopted will depend on the type of cheese to be produced. Enzymic coagulation is carried out at higher pH (5.8–6.5) and leads to a smoother, elastic precipitate which is easier to drain. The enzyme of choice for this process is chymosin which is extracted from the fourth stomach of unweaned calves. The important property of the enzyme is its ability to precipitate casein without promoting a high level of proteolysis. As the calves mature, chymosin is replaced by pepsin as the major stomach enzyme. Pepsin causes far more proteolysis of milk than chymosin and leads to the formation of short, bitter-tasting peptides which detract from the flavour of the finished cheese.

The suitability of chymosin for cheese manufacture, coupled with a limited source of supply, has led to significant shortages and elevated prices for this enzyme. To overcome this shortage, a range of alternative enzymes has been considered. As with chymosin, these should have a high clotting to proteolysis ratio. To date, the most successful replacement has been obtained from the mould *Mucor miehei*. This enzyme shows similar specificity to chymosin but is more stable and can lead to further, undesirable, protein breakdown after coagulation. Despite this disadvantage, commercial enzymes derived from *Mucor miehei* have been used successfully to replace chymosin in certain cheese manufacturing processes, without major changes in procedure being required. Although replacements for chymosin are successfully used, they are not suitable for manufacture of all cheese types. To overcome the shortage of chymosin, several biotechnology companies have undertaken genetic engineering programmes to produce chymosin from micro-organisms (see Chapter 2) (Beppu, 1983).

Other enzymes involved in the cheese production process include lipases, which are present in milk. These enzymes hydrolyse the fat component of milk, creating characteristic changes in taste. For some cheeses natural lipases can be augmented by the addition of extra enzyme. More generally, a range of other enzymes naturally present in milk can be significant in the modification of properties associated with the preparation of many milk products.

Proteolysis
The enzymic breakdown of protein is utilized in a number of industries for a variety of reasons, including changes in product taste, texture and appearance (Adler-Nissen, 1986). The plant-derived proteinases papain and bromelain are used as meat-tenderizing agents. Other microbial proteinases are used to break down the soluble proteins present in beer, which if left untreated would precipitate and form haze on cooling. This process is known as chill proofing. Other uses of proteinases for product modification include gelatin hydrolysis to prevent

subsequent gelation in food products, and the breakdown of gluten in bread dough to control viscosity during handling and improve final texture and appearance.

In addition to applications where the use of proteinases is peripheral to the main process, i.e. enhancing desired properties and reducing undesired effects, there are situations where the action of these enzymes is crucial to the whole process. The best example of this is soy protein hydrolysate and soy sauce manufacture (Yokotsuka, 1985). De-fatted soybean is widely used as a food material as it has a high protein content with a nutritionally balanced amino acid composition. However, in its natural state its structural properties are unsuitable for many food uses. To overcome this problem two types of soybean hydrolysates are produced: highly functional and highly soluble hydrolysates. The former retains some structure and is usually included in other food products in small amounts. The latter type may be incorporated at high concentrations to increase the protein levels of liquid-based foods. A special use of soy protein hydrolysis is the production of soy sauce. This is a fermentation process using *Aspergillus oryzea*, although it is the proteinases produced by the organism, and the resultant hydrolysis of protein, which is the major factor in the manufacture of soy sauce.

Carbohydrate degradation

The breakdown of starch to soluble sugars during the brewing process is central to the production of alcohol by traditional means (Godfrey, 1983). Historically, the breakdown of starch present in barley was initiated by encouraging germination. The growth associated with germination results in the synthesis of carbohydrate-degrading enzymes. Germination can be terminated by heating and drying, creating a crude enzyme preparation. Subsequent milling of the grain followed by soaking in warm water activates the enzymes which then continue to hydrolyse the starch. Although the major enzymes involved in the process are all present in the barley, it may be that the levels are not always optimal for good conversion. For this reason, malted barley is now commonly supplemented by the addition of exogenous enzyme (Table 7.1). Addition of supplementary enzymes allows the effects of grain quality variations to be evened out and allows the use of cheap alternatives to barley. Starch alternatives have been available for a long time but their lack of natural enzymes has restricted use in the brewing industry (Marshall *et al.*, 1982).

The most common form of carbohydrate encountered in biological systems is cellulose, with estimates of an annual world biosynthesis of 10^{11} tons. Obviously, this represents a vast potential feed-stock for a number of industries, and has created a great deal of research interest (Gaden *et al.*, 1976; Ward, 1985). Specific processes based on cellulase enzymes will be discussed later in the chapter, but there are a number of existing processes which can be enhanced by the addition of cellulase. Specific examples include silage production, alginate recovery and fruit processing.

Silage is the animal food product resulting from fermentation of grass. It is basically a preserving process where the pH of the material is lowered rapidly to prevent bacterial spoilage. The process can be accelerated by the addition of cellulases to liberate fermentable sugar and nitrogen products.

Table 7.1 Typical exogenous enzymes applied to brewing processes

Enzyme type	Beneficial action	Point of application
Bacterial α-amylases	Adjunct liquefaction	Decoction vessel (cereal cooker)
	Adjunct liquefaction	Mash vessel
	Malt improvement	
	Set mashes	Mash vessel
	Starch positive worts	Lauter or mash filter
Fungal α-amylases (maltogenic action)	Improved fermentability	Fermentation
	Low calorie and 'diet'	Fermentation
	Set mashes	Mash vessel
	Starch positive worts	Lauter or mash filter
Fungal amyloglucosidases	Low calorie and 'diet'	Fermentation
	Maximum fermentability	Fermentation
	Priming replacement	2° fermentation or post-pasteurization
Bacterial debranching enzymes	Maximum fermentability	Fermentation
Bacterial glucanases	Increased extract	Mash vessel
	Improved wort separation	Mash vessel
	Improved filtration	Mash vessel/fermentation/ conditioning tank
Fungal glucanases (including cellulases)	Improved extraction	Mash vessel
	Improved wort separation	Mash vessel
	Improved filtration	Mash vessel/fermentation/ conditioning tank
	Increased adjunct (especially sorghum)	Mash/decoction vessel
	Haze prevention	Mash vessel
	Haze removal	Fermentation/ conditioning tank
Bacterial neutral proteinase	Increased adjunct	Mash vessel
	Nitrogen regulation	Mash vessel/fermentation
Plant proteinases (papain)	Chillproofing against protein hazes	Conditioning tank
Fungal pentosanases	Prevention/removal of specific haze components	Mash vessel/fermentation/ conditioning tank
	Improved extract (especially wheat and sorghum)	Mash vessel

(From Godfrey, 1983)

Alginates are polysaccharides that gel in the presence of divalent cations, e.g. Ca^{2+}, which have important applications in a range of industries. At present the major source of alginates is the brown seaweeds. Cell-wall-degrading enzymes break down tissue, allowing improved extraction of alginate.

Cellulases also find application in the solubilization of fruits in the manufacture of juices. In related processes they are used for the extraction of pigments and pectins from fruit skins.

Sugar refining

The extraction of sucrose from sugar beet molasses can be complicated by the presence of raffinose, a trisaccharide which prevents crystallization. To increase the sugar yield and improve process throughput, the raffinose can be degraded enzymically. The result of this degradation is both to improve the crystallization and to produce sucrose as one of the products of the hydrolysis. The enzyme α-galactosidase is produced by the mould *Morteirella vinaceae raffinosutilizer* and can be conveniently used by immobilizing the mycelial pellets that this organism produces. The hydrolysis reaction is carred out at a pH above 5 to avoid the acid-catalysed inversion of sucrose. A similar approach is sometimes required in the processing of cane sugar, where starch is broken down prior to crystallization using α-amylase (Park *et al.*, 1983).

Waste treatment

In considering the applications of enzymes in waste treatment, a distinction must be made between situations where the waste from one process is the raw material of the next, e.g. starch conversion, and processes aimed solely at reducing the associated cost of treatment. There are a large number of food-processing industries which produce wastes requiring treatment (see Table 7.2).

The applications of enzymes stem from the need to break down complex polymers to enhance their subsequent microbial degradation. Examples include the use of lipases in association with bacterial cultures to remove fat deposits from the walls of effluent-carrying pipes. Other polymer-degrading enzymes used in this way are cellulases, proteinases and amylases. One particular application which can be described loosely as waste treatment, is the use of proteinases in commercial detergent preparations, the so-called biological washing powders (Barfoed, 1983).

In addition to the specific breakdown of polymeric material, there are also applications for enzymes capable of degrading compounds of high toxicity which would inhibit microbially based treatment processes. A specific example which has been reported is the use of horseradish peroxidase to initiate the degradation of phenols and aromatic amines which occur in many industrial waste waters (Klibanov *et al.*, 1983). In the longer term it is anticipated that processes based on organisms genetically engineered to degrade these compounds will represent a more economic approach.

Development of novel processes

There has been a great deal of interest in the use of enzymes for the production of novel food materials or developing new routes to existing processes. The majority of these developments have been focused on the area of carbohydrate processing. Carbohydrate wastes such as starch, cellulose and lactose arise from a range of industries. As raw materials these can potentially be converted into a range of

Table 7.2 Major waste creating industries producing bioprocess wastes

Waste material type	Industrial activity
Starch	Bread, flour, confectionery
	Brewing
Cereal	Cereal foods
	Distilling
Sugars	Finished compound foods
	Food ingredients
	Paper, adhesives
	Sweeteners
	Textiles
Cellulose	Paper, timber
Lignocellulose	Brewing, distilling
Proteins	Abattoir
	Butchery
	Cereal extraction
	Dairy
	Poultry
	Finished compound foods
	Brewing, distilling
	Fish processing
	Vegetable processing
	Leather
	Gelatin
	Single-cell fermentation
	Oil seeds processing
Fats	Abattoir
	Butchery
Oils	Poultry
	Dairy
	Fish processing
	Oil seeds processing
	Finished compound foods
	Cereal foods

(From Godfrey, 1983)

higher value products. Unlike the pharmaceutical markets where drugs sell for thousands of pounds per kilogram, the product from carbohydrate-degrading processes will be priced in terms of pence per kilogram and so industries based on these conversions must concentrate on high-throughput, low-cost processes. In terms of process engineering, these applications represent the most highly developed applications of enzymes.

Cellulose degradation
In addition to the uses of cellulose previously outlined, there is a great deal of interest in the use of these enzymes to degrade cellulose prior to an ethanol-producing fermentation. The problems that limit this approach at present stem from the structural arrangement of cellulose in the native state. Currently, the most effective enzyme preparation is obtained from the fungus *Trichoderma reesei*. The catalyst preparation represents a mixture of cellulase and cellobiase activities. Lignin is also present in the complex, but is unaffected by commercial cellulase products. Unfortunately, the cellobiase activity in commercial preparations is low and is further inhibited by the product of hydrolysis (glucose). To overcome this problem, supplementary cellobiase may be added from another source.

Before enzyme degradation can be used successfully, a number of pre-treatments are necessary to convert the cellulose to an accessible form. These include milling, and extraction of lignin with hot solutions of higher alcohols. Other techniques which have been used with some success are chemical swelling with alkali, and steam explosion where the cellulose fibres are saturated with steam under pressure which is then rapidly released, causing the fibres to explode. Although no process is yet operated commercially, a pilot plant has been constructed to convert one ton of cellulose-based compounds per day using simultaneous enzyme saccharification and fermentation to produce ethanol. This has yet to be scaled up to an industrial scale operation, possibly for economic rather than technical reasons.

Lactose hydrolysis
Lactose is the major carbohydrate of milk and is present at a concentration of approximately 4% (w/w). Large quantities of lactose-containing whey are produced during the manufacture of cheese. This material contains an extremely high biological oxygen demand (BOD) load and is costly to treat as a waste. Consequently a number of processes have been developed to utilize the organic components of whey. The first stage in such a process is usually to recover the protein component remaining after cheese manufacture. After the protein has been removed, a solution containing lactose and various salts remains. While lactose has a variety of uses, including animal feed, pharmaceutical product formulation and fermentation feed-stock, its limited digestibility and low sweetness value have restricted its applications. Hydrolysis of lactose to its component monosaccharides (D-galactose and D-glucose) improves the solubility, digestibility and sweetness of the product and so represents a desirable conversion of a low-value feed-stock (Gekuas and López-Leiva, 1985).

While chemical hydrolysis of lactose is possible, there is increasing interest in the use of the enzyme β-galactosidase to effect conversion. For industrial use the enzyme is produced from microbial sources. The stability and activity of the enzyme depends on the source and for safety reasons the β-galactosidase is usually extracted from the food yeasts *Klyveromyces fragilis* or *Klyveromyces lactis*. Applications have been reported using both soluble and immobilized β-galactosidase and at present there are two large-scale processes in operation based on immobilized enzymes.

Starch conversion
Processes producing sweeteners based on starch degradation have been used for many years, although originally these were based on acid-catalysed hydrolysis. During the course of development and expansion of the starch conversion industry, there has been a switch from acid catalysis to the use of enzymes. The high specificity of these enzyme reactions, coupled with the high activity of the preparations available, means that throughput can be increased and the formation of undesirable byproducts minimized (Coker and Venkatasubramanian, 1985).

By far the most successful application of enzymes on an industrial scale is the production of high fructose corn syrup (HFCS). The aim of this process is to produce a material of equivalent sweetness to sucrose from a low cost raw material (corn starch) (Antrim, 1980). This conversion requires the sequential use of three enzymes. First, the starch (a polymer of D-glucose) is partially degraded, using a bacterial α-amylase. This material is further degraded to give a solution where 94–96% of the carbohydrate present is in the form of D-glucose. Glucose has some applications in the food and pharmaceutical industries, but its use is restricted by its limited solubility at high concentrations and its low sweetening power compared with sucrose. For these reasons the bulk of glucose produced is converted to a mixture of glucose and fructose using the enzyme glucose isomerase.

The initial interest in HFCS was stimulated by high sugar prices on the world market. However, progress was impeded for some time by the lack of a suitable isomerase enzyme. Native glucose isomerase has a requirement for stoichiometric quantities of an expensive coenzyme (ATP), making it unsuitable for the large-scale process envisaged. In the 1950s it was found that naturally occurring xylose isomerases were capable of isomerizing glucose under some conditions. Further work related catalytic activity changes to substitution of essential metal ions. These results allowed the modification of bacterial D-xylose isomerases to allow the isomerization of D-glucose (Chapter 9). While the α-amylase and glucoamylase enzymes are naturally extracellular, the isomerase is an intracellular enzyme. It naturally has a lower catalytic activity and a lower thermal stability. So, while it is common practice to carry out the starch degradation batchwise using soluble enzymes, the isomerization is usually achieved in a packed-bed reactor containing immobilized catalyst. There are a number of commercially available isomerase preparations (some based on heat-fixed immobilized cells, others on extracted immobilized enzymes), clearly demonstrating that there is no ultimate criterion for deciding between immobilized cells and enzymes.

The raw material (corn) is milled and the resultant starch grains suspended to give a 30–35% (w/v) slurry. This material is obviously difficult to handle (having a high viscosity and suspended particulate material), therefore the next stage in the process is to liquefy the suspended starch. This is achieved using a bacterial α-amylase. The enzyme is mixed with the starch slurry and held at a temperature of 80–110 °C for a period of 2–4 hours at pH 6–6.5. At this stage calcium ions are added to activate the α-amylase enzyme. The α-amylase enzyme catalyses the hydrolysis of α1–4-linked glucose units but is incapable of breaking branched chains, and so degradation is limited by the amount of branching in the chain. The resultant 'limit dextrin' material has to be broken down using a second enzyme, pullulanase. Prior to addition of glucoamylase, liquefied starch solution must be cooled to 60 °C and the pH adjusted to 4–5. The holding time for the material in the reactor at this stage will be between 24 and 90 hours, depending on the throughput time required and hence the amount of enzyme added. At the end of this time, a product concentration of between 94–96% dextrose is required for a viable isomerization stage.

As HFCS is used as a food material, the purity criteria governing its composition are stringent. For these to be met the saccharified material has to be cleaned up prior to the isomerization stage. A clean feed-stock is also important if optimum catalyst life is to be achieved in the immobilized isomerase reactor. The cleaning process involves filtration to remove precipitated proteins and particulate material. Although enzyme hydrolysis minimizes the formation of coloured byproducts, an active carbon adsorption method is used to remove the small quantities still produced. The pH optimum for the isomerization reaction lies between 7.5 and 8.2, and so the pH has to be raised from the value of 4–5 used during the second stage of saccharification. These pH adjustments tend to elevate the inorganic salt concentration in the liquor, and so deionization is required at this stage in the process. Finally, the solid concentration is adjusted by evaporation and magnesium ions are added as an activator.

The feed-stock is fed to the isomerase reactor at a solids' concentration of 40–45%. The equilibrium constant for the reaction is unity at 60 °C, but for a realistic residence time the conversion is limited to give 42% fructose. Initially, this was the only product on the market, but for some applications a higher fructose content is required. More recent developments in adsorption chromatography have allowed selective concentration of the fructose such that a solution containing 90% of the carbohydrate present as fructose can be produced. This material, referred to as very enriched fructose corn syrup (VEFCS), can be blended with the 42% product of the isomerization reaction to give the 55% fructose solution which is ideal as a commercial sweetener. The overall production scheme is outlined in Fig. 7.1.

Enzymic production of amino acids
The production of amino acids is well suited to enzyme technology. Although they can be synthesized using a purely chemical approach (Wiseman, 1985), this leads to the formation of a racemic mixture of D and L isomers. As it is only the L isomer which is biologically active, the mixture must be resolved into its component forms. This resolution can be achieved using the enzyme aminoacylase. Once synthesized, the DL mixture of amino acids is first acetylated. Treatment of the

Enzymes in agriculture and the food industry

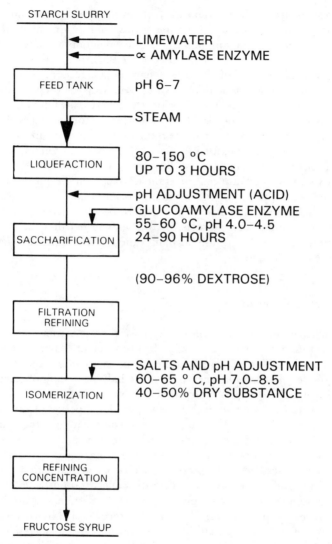

Fig. 7.1 A process sequence for starch liquefaction and saccharification. (From Antrum, *et al* 1979)

mixture with aminoacylase leads to the selective removal of the acetyl group from the L-isomer, which can then be separated. This process has been operated for about ten years in Japan, and represents the first example of the use of an immobilized enzyme on an industrial scale. The scale of the process is such that a 1 m^3 reactor is used to give a quantitative conversion of the L isomer component. The productivity varies with the amino acid used, but lies between 4.5 and 10.5 tons per column loading of enzyme. The operating temperature for the conversion is 50 °C, giving a catalyst half-life of 65 days.

Going a stage further, it is possible to synthesize enzymically amino acids from organic acid precursors. A major example of this approach is the production of aspartic acid from ammonium fumarate. This reaction is catalysed by the aspartase activity present in immobilized *E. coli* cells. Although the whole cells are entrapped, typically in beads of the polysaccharide carrageenan, some elimination of unwanted side reactions is achieved by heat shocking the preparation prior to use. This process has been operated since the early 1970s, the immobilized system being approximately 60% cheaper than earlier fermentation processes.

Other important amino acids whose production involves an enzyme-mediated step, include D-phenylglycine, used in the synthesis of semi-synthetic penicillins, and L-tryptophan, an essential amino acid which can be synthesized from indole. It is these two areas which are central to the development of this technology. In terms of large-scale application, the production of essential amino acids as dietary supplements is particular important. If single-cell protein becomes established in the animal and human food markets, it can be expected that the demand for essential amino acids will increase, as many microbial proteins are deficient in some of these crucial residues.

Economic considerations

Unlike pharmaceutical applications of enzymes, the use of enzymes in the food industry requires that a low-value product be produced at a lower cost than other existing alternatives. The enzymes used in this process can be regarded as being unusually stable, for example, the operational half-life of immobilized glucose isomerase is between 70 and 120 days at 60 °C. Normally the preparation would be used for a period of two half-lives, i.e. an operational period of approximately nine months (Daniels, 1985). Originally the cost contribution of the immobilized enzyme to the product was estimated at US $14.00 per ton.

For those conversant with the laboratory use of enzymes, it is clear that the conditions used at all stages in the production of HFCS are exceptional. It is usual in biochemical studies to use low substrate concentrations and to monitor reactions at ambient (25 °C) or physiological (37 °C temperatures). For an economic process to produce HFCS, current processes are based on substrate concentrations of 50% (w/v), quantitative conversions and solids residence times in the packed-bed reactor of the order of 0.5–5 minutes. Notwithstanding these constraints, the glucose isomerase preparation must be expected to have an operational half-life of better than 30 days. When considered in this light it becomes obvious why there are only a few full-scale processes based on enzyme technology. However, research is continuing to improve techniques for enzyme modification and stabilization (Chapters 6 and 9). In cases where substrate and product solubility are the limiting factors in the process, the use of non-aqueous solvents may well be applicable, and this is an active area of academic and commercial research (see Chapter 10).

Finally, one of the major restrictions for the use of enzymes is the requirement for high-cost enzyme cofactors for many synthetic reactions. While work is still continuing to solve the problems of coenzyme regeneration, this still represents a major bottleneck. However, it is interesting to note that a West German company have introduced recently a process for producing amino acids which requires the coenzyme NADH as a substrate. The reactor incorporates an ultrafiltration membrane which retains NADH coupled to a soluble polymer. The NADH can be regenerated *in situ* using formate dehydrogenase (Wandrey and Wichmann, 1985). Although many coenzyme regenerating systems have been demonstrated on a laboratory scale, this represents the first commercial-scale process.

Chapter 8

Enzyme-based sensors

Introduction

One of the earliest areas of enzyme application was in the field of analysis. Many compounds cannot be detected directly using existing techniques and so it is often necessary to convert them to products whose levels can be quantified. There are a number of colorimetric chemical assays which involve a prior conversion step. The use of enzymes in analysis is simply a logical extension of this approach.

The assay of a particular compound in a complex biological fluid such as plasma is greatly complicated by the presence of a number of potentially interfering substances. One approach to solving this problem is to use an appropriate separation technique coupled with a sensitive but non-specific detector. This approach is perhaps best illustrated with the example of high performance liquid chromatography which is now used in a wide range of biological and pharmaceutical assays. The alternative is to use a specific detector which is not affected by the presence of contaminating compounds. The biological activity and specificity of enzymes make them obvious candidates for this type of detector.

The simplest approach is to carry out a direct enzyme-based assay, i.e. use a suitable enzyme to convert a compound which cannot be easily detected to give a product or byproduct which can be. Although there are many cases where this cannot be accomplished by a single reaction, it is possible to link a sequence of reactions together such that the product of the terminal reaction can be measured. For experimental convenience this is often accomplished by coupling the reaction to the oxidation or reduction of the coenzyme nicotinamide adenine dinucleotide (NAD(P)(H)). The oxidation or reduction of this compound is accompanied by a large change in absorbance at a wavelength of 340 nm (Fig. 8.1).

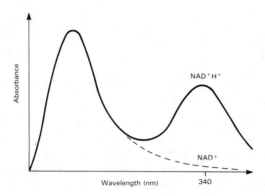

Fig. 8.1 Absorbance against wavelength scan of NAD^+/NADH.

The molar extinction coefficient for NADH at a wavelength of 340 nm is 6.22×10^3 l mol^{-1}. It can be calculated that conversion of one micromole of NAD^+ to NADH in a 3 ml cuvette with a 1 cm light path would cause an absorbance change of greater than 2 units. With modern spectrophotometers this means that concentrations of the order of nanomoles per litre can be measured.

This approach has been used for a number of compounds of medical interest.

Ammonia

$$\text{2-oxoglutarate} + NH_4^+ + NADH \xrightarrow{\text{glutamate dehydrogenase}} \text{glutamate} + NAD^+ + H^+$$

Glucose

$$\text{glucose} + ATP \xrightarrow{\text{hexokinase}} \text{glucose-6-phosphate} + ADP$$

$$\text{glucose-6-phosphate} + NADP^+ \xrightarrow{\text{glucose-6-phosphate dehydrogenase}} \text{6-phosphogluconate} + NADPH$$

Enzyme analyses like these have been used for many years and the trend is continuing as a result of the increasing availability of highly purified enzymes. A comprehensive handbook of enzyme based analyses is edited by Bergmeyer (1986). Analyses of this type have an important role in analytical chemistry but they suffer from being costly and time-consuming. For applications which require routine analysis, some form of automation, coupled with re-use of the expensive enzyme, is required.

Immobilized enzymes

Developments in enzyme immobilization techniques have made a significant input into the area of enzyme-based sensors. The advantages of immobilized enzymes in analysis can be summarized (Bowers and Carr, 1980) as follows:

(1) increased stability
(2) catalyst may be re-used
(3) the sample can be separated from the enzyme for further assays
(4) predictable activity decay curves

In addition, the use of multi-enzyme systems may allow unstable reagents to be generated *in situ* to avoid problems of instability.

The use of immobilized enzymes in an analytical system can be divided into two categories. First, an immobilized enzyme reactor can be constructed which produces one or more products capable of detection by existing methods (e.g. absorbance change). Second, the enzyme can be immobilized around, or onto, a transducer which is capable of detecting physical or chemical changes in its immediate surroundings.

Analytical reactors

These can be considered as an example of flow injection analysis. Basically the system is similar to liquid chromatography but without a separation stage (Fig. 8.2) (Pedersen and Horváth, 1981). The reactor would normally be one of two types.

(i) *Packed-bed reactor*. The enzyme is immobilized to a particulate support which is packed into a column. The liquid containing the sample is passed through the column, allowing the reaction to occur. The advantage of this system is that there is a large surface area for immobilization, so high enzyme loading is possible. As the degree of conversion is a function of enzyme loading and throughput, the higher the enzyme loading, the faster the sample throughput can be. However, problems arise from dispersion of sample as it passes through the column. This

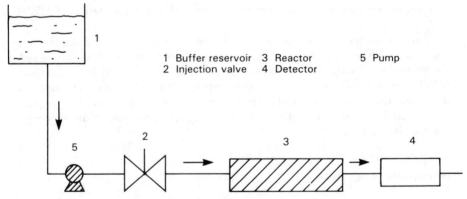

Fig. 8.2 Apparatus layout for flow injection analysis.

causes dilution of the sample/products and hence tends to reduce the sensitivity of detection. If the output of the detector is considered as a peak on a chart recorder trace, then dispersion would lower the height and increase its width. The broader the responses, the longer the period which must be allowed between samples to avoid cross-contamination. Therefore, while packed-bed systems offer greater activity for continuous monitoring, they have some drawbacks for assays which have to be performed on a large number of discrete samples.

(ii) *Open-tubular reactors.* The enzyme is immobilized onto the inside of a length of tubing, typical internal diameter 1 mm. The sample is then passed through this length of tubing, which serves as the reactor. The disadvantage is that there is a relatively small surface area for immobilization if the tube length is to be kept to a practical length (< 10 m). In practice, this reactor system has received the most analytical interest because of its suitability for use in segmented flow analysis. Traditionally, these devices have been used extensively for analysis in the clinical, environmental and pharmaceutical fields. This type of reactor is especially suited for reactions requiring long incubation times while maintaining precision. As each sample enters the system as a liquid slug, axial dispersion is limited and potential sample overlap is avoided. Although the low surface area is a potential problem, studies of nylon tube reactors have shown that this is offset by the higher mass transfer rates obtained with tubular reactors.

Analytical enzyme reactors have been used in a wide range of determinations. The earliest, reported in 1966, was for the detection of lactate and glucose. Further development led to a method published in 1976 for the detection of nitrate. The advantages of this method led to its adoption as an official analysis technique in the USA.

Transducer-bound enzymes

Although widely used, enzyme reactor based analysers are complex and expensive pieces of equipment. An alternative is to associate the enzyme with the detector to simplify the operation. The aim is to produce a cheap robust sensor which is capable of giving a continuous output. In an ideal case, such a sensor would require no external reactants other than the substrates, and would be as non-destructive as possible, i.e. only removing a small fraction of the substrate.

The simplest form of enzyme-transducer sensor is the so-called enzyme electrode (Danielsson, 1985). In this case, the enzyme is bound to an ion-selective electrode which will detect the presence of a substrate or a product of the enzyme-catalysed reaction. An example of this approach would be the association of urease with an ammonium ion probe such that the concentration of an urea solution could be measured from the ammonia released as the urea was hydrolysed (Fig. 8.3).

A large number of enzyme electrodes have been based on the oxygen probe in conjunction with enzymic redox reactions. An example of this is the detection of glucose with the enzyme glucose oxidase (see Chapter 5). As the reaction proceeds the oxygen concentration falls; this change can be monitored continuously with

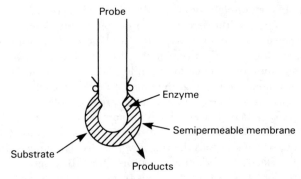

Fig. 8.3 A simple enzyme electrode.

the probe. In practice, this response will be modified by a number of external factors that will influence the time taken for the probe to respond. The measurement is usually taken after the probe has reached a steady state, i.e. when the rate of substrate supply to the enzyme and the rate of substrate consumption are equal. In practice, this point would be reached between 30 seconds and 10 minutes.

The sensing probes used for enzyme electrodes fall into two groups: potentiometric and amperometric. Potentiometric probes, e.g. the glass-pH electrode, determine the potential between a reference and a sensing electrode (Williams and Wilson, 1975). The structure of a combined probe is shown in Fig. 8.4. The system represents an electrical cell with the observed potential being a result of the sum of three components:

(1) the internal reference electrode
(2) the asymmetric potential
(3) the potential due to the different concentration of hydrogen ions on each side of the membrane

The first two components are constants for the electrode, leaving the concentration of hydrogen ions as a variable which is thought to result either from the transfer of hydrogen ions through the glass or from an ion-exchange process. The probe's response in terms of pH can be quantified as

$$E_G' = E_{ref} + E_{asym} + 0.059\,(pH_x - pH_c) \quad (25\,°C)$$

where E_G' = measured potential of the glass electrode
E_{ref} = potential of the internal reference electrode
E_{asym} = asymmetric potential
pH_c = pH of the internal solution of the glass electrode (constant)
pH_x = pH of the unknown solution
Therefore

$$pH_x = \frac{E_G' - E_{constant}}{0.059}$$

Enzyme-based sensors 107

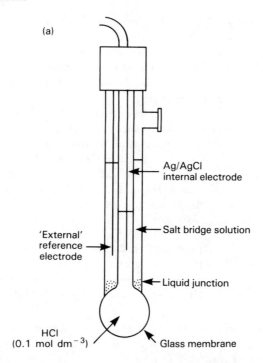

Fig. 8.4 (a) Structure of a combination pH probe. (b) Half-cell reactions of a combination probe.

The response of the probe will change with temperature as described by

$$E_G' = E_G - \frac{2.303\,RT}{F}\text{pH}$$

where E_G' = probe potential
E_G = standard potential
R = gas constant
F = Faraday constant
T = absolute temperature

As pH is known to affect enzyme kinetics (Chapter 4), it is apparent that this is not an ideal parameter to measure. As the pH changes due to the enzyme-catalysed reaction, so will the catalytic activity of the enzyme, thus causing a non-linear response of the probe to concentration. However, by modifying the properties of the glass membrane, electrodes of this type can be made selective to a number of ions, e.g. NH_4^+ and K^+, which are more suitable for use in an enzyme probe. It is apparent from theoretical considerations that the response of an enzyme electrode will change in response to external temperature changes. The response of the glass electrode is well characterized and most instruments will allow for temperature compensation. However, the effects on enzyme activity and stability have also to be considered.

The second category of electrodes is the amperometric type. The most common example of this type of probe is the Clark electrode for the determination of oxygen concentration. This is based on a platinum cathode and silver anode immersed in the same solution of potassium chloride but separated from the test solutions by a polytetrafluoroethylene (PTFE) membrane. A potential of 0·5–0·8 V is applied across these electrodes and the current generated is proportional to the concentration of substrate in the test solution. The reactions that occur in an oxygen electrode are as follows:

$$\begin{array}{ll} \text{Anode} & 4Ag + 4Cl^- \longrightarrow 4AgCl + 4e^- \\ \text{Cathode} & 4H^+ + 4e^- + O_2 \longrightarrow 2H_2O \\ \hline & 4H^+ + 4Ag + 4Cl^- + O_2 \longrightarrow 4AgCl + 2H_2O \end{array}$$

The difference in operation of the two types of probes has implications for the performance of an enzyme electrode based on that particular mode of operation. In the case of potentiometric probes, the potential is measured using a high impedance circuit; hence there is essentially no current used, and so no removal of substrate or product occurs. In the case of the amperometric oxygen electrode, the substrate of the enzyme reaction is consumed also by the reactions taking place in the probe.

The rate processes which can modify the behaviour of the probe are

$$\text{Bulk diffusion}$$
$$[R] \longrightarrow [R]_{\text{electrode surface}}$$
$$\text{Internal diffusion}$$
$$[R]_{\text{electrode surface}} \longrightarrow [R]_{\text{enzyme}}$$

These describe transport of the reactant. Removal of the reactant can be described in terms of the enzyme kinetics and, in the case of amperometric probes, the reaction at the probe surface. The effects of diffusion resistance can be modelled in terms of mass transfer with chemical reaction in a similar way to entrapped enzyme reactors (Chapter 6). In practice the factors affecting a probe's suitability are:

(1) response time
(2) wash time
(3) linear range

As previously discussed, the response time is reflected by the time taken for the

Enzyme-based sensors

mass transfer to be balanced by catalytic removal. The wash time is the time taken for the probe output to return to its base level when placed in a solution free of reactant. The linear range of the probe is dictated by the K_m of the enzyme. At reaction concentrations significantly below K_m, the response of the probe will be first order and therefore linear. As concentrations are increased and the enzyme is saturated, the rate becomes zero order and non-linearity results. However, if mass transfer resistance is significant, then the substrate concentration in the immobilized enzyme layer will be lower than the bulk solution, and therefore the effective linear range will be extended. However, although mass transfer effects may increase the linear range, this advantage will be offset by longer response and wash times, which will limit the rate of sample determination.

While the amperometric oxygen probe has been most widely used as the basis of enzyme electrodes, the linear concentration range is also constrained by the limited solubility of oxygen. To overcome this problem other electron acceptors of greater solubility can be used, e.g. benzoquinone (Williams *et al.*, 1970). Some examples of enzyme electrodes are shown in Table 8.1.

Although enzyme probes have been used successfully, it is probably fair to state that their development has been overshadowed by advances in other areas of enzyme sensor development. Their chief advantages can be assigned to their relatively robust nature, ease of construction and the simplicity of associated equipment.

Enzyme thermistors

This technique has developed from work in microcalorimetry. Early systems were analogous to the enzyme reactor analysis discussed previously. In this case an exothermic reaction is catalysed by an immobilized enzyme and the heat evolution calculated by measuring the temperature change of the reaction liquid. These sensors were very sensitive but had a slow response. The development of thermistor devices based on semiconductor materials showing a large resistance change with respect to temperature, allowed the miniaturization of the enzyme-based microcalorimeters and led to the concept of the enzyme thermistor (Mosbach and Danielsson, 1981). In this device a well-insulated, immobilized-enzyme column is constructed, and a thermistor is mounted in the centre of the column-packing material. This transducer is sensitive to extremely small changes in temperature which are reflected in a changed impedance. Studies show that approximately half of the heat evolved as a result of the enzyme reaction can be registered as a temperature change. Changes between $0.004\,°C$ and $1.0\,°C$ are typically observed and lie within the sensitivity range of the detector.

Typical designs of sensor housing can be seen in Fig. 8.5a showing both a simple single thermistor device, and a system employing a reference thermistor (Fig. 8.5b). The main experimental problem associated with the use of enzyme thermistors is the need to minimize fluctuations in external temperature. The use of water baths and insulated housings is often sufficient; however, in some cases the use of a reference thermistor allows greater sensitivity.

Table 8.1 Examples of enzyme electrodes based on amperometric and potentiometric probes

Substrate	Enzyme	Electrode	Response time	Linearity	Stability
Alcohols	Alcohol oxidase	Pt (O_2) −0.6 V vs SCE	~2 min	Up to 5 mg% Ethanol	~120 d
		Pt (H_2O_2) +0.6 V vs SCE	~2 min	$1-25 \times 10^{-9}$ M Methanol	~24 h
	Alcohol dehydrogenase	Pt (NADH)	—	0–1 m nonlinear	~8 samples
L-Amino acids	L-Amino acid oxidase	Pt (H_2O_2) potential scan	30–60 s	$1-400 \times 10^{-6}$ M	10–12 d
		Pt (O_2) −0.6 V vs SCE	1–2 min	$0.1-1 \times 10^{-3}$ M Phenylalanine	> 120 d
Glucose	Glucose oxidase	Pt (H_2O_2) +0.6 V vs SCE	1 min (steady state)	$0.5-15 \times 10^{-3}$ M	300 d
		Pt	3–10 min	$1-20 \times 10^{-3}$ M	—
		Clark O_2 electrode	—	0.1–2.0 mg l^{-1}	—
Diamines	Diamine oxidase	Pt (O_2) −0.6 V vs SCE	<15 s	$20-400 \times 10^{-6}$ M	25% decrease in 14 d

Table 8.1 *(cont)*

Analyte	Enzyme	Electrode	Response time	Range	Stability
Lactate	Lactate dehydrogenase	Pt (Fe(CN)$_6$) +0.25 V vs SCE	~1 min	0.1–5×10^{-3} M	—
		Glassy carbon (NADH) +0.7 V vs SCE	3–4 min	0.1–4×10^{-3} M non-linear	Decrease of 20% in 2 h
Uric acid	Uric acid oxidase	Pt (O$_2$) −0.6 V vs SCE	2–3 min	0.1–1 mg l^{-1}	~100 d
Phosphate	Alkaline phosphatase/ glucose oxidase	Pt (O$_2$) −0.6 V vs SCE	1–2 min	1–10×10^{-3} M	~90 d
Urea	Urease	Beckman univalent cation	25–60 s for 98%	0.02–8 mg l^{-1}	>14 d
			60–90 s	10^{-4}–10^{-2} M	>21 d
		Instrumentation laboratories CO$_2$ electrode	1–6 min	10^{-4}–10^{-2} M	>3 d
		Nonactin in silicon rubber	60–80 s	3×10^{-5}–3×10^{-3} M	>7 d
		Air gap	1 min @ >10^{-2} M 4 min @ <10^{-3} M	10^{-4}–2×10^{-2} M	>21 d (300 assays)

Table 8.1 *(cont)*

		NH$_3$ gas sensor	1.5–2 min @ 10^{-2} M pH 9	5×10^{-4}–5×10^{-2} M	~20 d
Glucose	Glucose oxidase	pH glass	—	5×10^{-5}–5×10^{-3} M	>14 d
		Iodide ion selective			
		pH electrode	—	10^{-3}–10^{-1} M	>14 d
Amygdalin	β-Glucosidase	Heterogeneous	5 min	10^{-4}–10^{-1} M	~7 d
		CN$^-$ solid state electrode	<1 min	5×10^{-7}–5×10^{-3} M	>14 d
Penicillin	Penicillinase	pH glass	>2 min	10^{-3}–10^{-2} M	>7 d
		pH glass			
Creatinine	Creatininase	NH$_3$ gas sensor	6–10 min @ <5×10^{-3} M	7×10^{-5}–10^{-2} M	~4
			2–5 min @ >5×10^{-3} M		
L-Amino acids	L-Amino acid oxidase	Beckman univalent cation	1–2 min	10^{-4}–10^{-3} M	14 d
Uric acid	Uric acid oxidase	CO$_2$ gas sensor	5–15 min	2×10^{-4}–3×10^{-3} M	~10 d

(From Bowers and Carr, 1980)
SCE is the Standard Calomel Electrode

Enzyme-based sensors

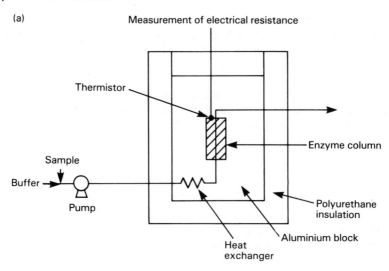

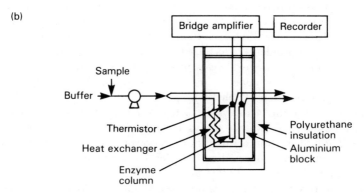

Fig. 8.5 (a) Enzyme thermistor housing using a single sensing thermistor. (b) Layout of a dual thermistor system.
(From Mosbach and Danielsson, 1981)

The response of a thermistor to a change in temperature can be described by

$$R_2 = R_1 \exp\left(\frac{B}{T_2} - \frac{B}{T_1}\right)$$

where B = characteristic temperature constant
T_1, T_2 = thermistor temperature
R_1 = resistance of thermistor at T_1
R_2 = resistance of thermistor at T_2

So the change in temperature of the thermistor can be calculated from its resistance change. The heat capacity of the immobilized enzyme column can be determined by including a wire of known resistance in the bed and passing a constant electrical current through it; the relationship between the energy input and the temperature change gives the heat capacity.

In the case of a system employing a reference thermistor, the simplest design is to incorporate the two thermistors into a differential bridge (Fig. 8.6) (Hubble, 1986). In this case a temperature change affecting only one thermistor will be recorded as a change in bridge output. Changes affecting both thermistors will not unbalance the bridge and so the output voltage will not alter. This can be described theoretically as

$$v = \left(\frac{R_1}{R_1 + R_2} - \frac{R_1 + \delta R}{R_1 + \delta R + R_2} \right) \cdot V$$

where v = change in output voltage
 R_1 = resistance of sensing thermistor
 R_2 = resistance of reference thermistor
 V = excitation voltage
 δR = change in resistance of sensing thermistor

From this relationship it is obvious that the higher the excitation voltage, the higher the sensitivity of the circuit. This is true up to the maximum power rating of the thermistor. However, at high excitation voltages, power dissipation from the thermistor can cause problems of localized heating in the column.

Enzyme thermistor systems have been used for a wide range of assays but obviously their suitability depends to some extent on the enthalpy change associated with the reaction (Table 8.2).

In addition to the direct assays, i.e. where the enzyme is used to detect its substrate, a number of indirect assays have been developed using enzyme thermistors. The best example of this is the detection of pesticides by their ability to inhibit the enzyme acetyl cholinesterase. In this case, a relatively high substrate concentration is included in the buffer, giving a constant temperature reading. As pesticide samples pass through the reactor, the reaction rate is depressed and so the

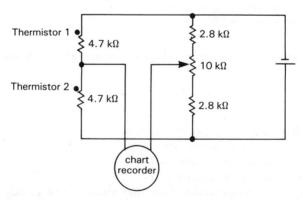

Fig. 8.6 Example of a bridge circuit for use with a double thermistor device.

Table 8.2 Molar enthalpies of some enzyme-catalysed reactions

Enzyme	EC number	Substrate	$-\Delta H$ (kJ mol^{-1})
Catalase	1.11.1.6	Hydrogen peroxide	100.4
Cholesterol oxidase	1.1.3.6	Cholesterol	52.9
Glucose oxidase	1.1.3.4	Glucose	80.0
Hexokinase	2.7.1.1	Glucose	27.6
Lactate dehydrogenase	1.1.1.27	Na-pyruvate	62.1
Trypsin	3.4.21.4	Benzoyl-L-arginine amide	27.8
Urease	3.5.1.5	Urea	6.6
Uricase	1.7.3.3	Urate	49.1

(From Mosbach and Danielsson, 1981)

temperature falls. Another interesting adaptation is the thermal enzyme-linked immunosorbent assay (TELISA). In principle, this technique is similar to the ELISA protocol outlined in Chapter 5. For example, an antibody to a drug is immobilized in the reaction column and the sample (antigen) is applied with a resultant blocking of a proportion of the antibody sites. A portion of the enzyme–drug conjugate is then passed through the column, interacting with unbound antibody. When substrate is passed through the column, the temperature change will be a function of the amount of enzyme bound and hence the amount of drug present in the sample.

Enzyme thermistors have been described mainly with respect to assays of discrete samples. Their sensitivity (of the order of 10^{-5} mol l^{-1} is rather limited but, as they rely on a general detection principle which is not affected by solution turbidity, they have a wide range of potential applications. Although limited baseline stability restricts their application for continuous analysis, enzyme thermistors have been proposed for continuous monitoring of bioreactor output. A system has been demonstrated which uses an enzyme thermistor to sense changes in effluent concentration, and to control the feed pump in an immobilized enzyme reactor hydrolysing lactose to D-glucose and D-galactose (Danielsson *et al.*, 1979).

Enzyme field effect transistors

These devices (usually called ENFETs) are analogous to the enzyme electrodes discussed earlier. The ion-selective field effect transistor (ISFET) can be regarded as an amplifier (Fig. 8.7). The amount of current passing through the device is proportional to the concentration of an external ion. These devices are constructed using integrated circuit technology and so are amenable to mass production (Moss *et al.*, 1978). The advantages of ISFETs over ion-selective electrodes can be summarized:

(1) small solid state construction allows increased reliability and ruggedness

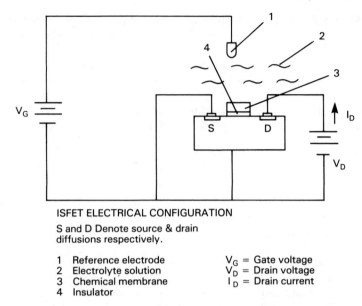

Fig. 8.7 Structure of an ion-selective field effect transistor. (From Moss et al., 1978)

(2) rapid response time arising from the small detector area and the thin membrane used
(3) possibility of producing integrated circuits containing multiple ISFETs of differing specificity
(4) ease of incorporation into associated monitoring circuitry

As with an ion-selective probe, membrane properties can be varied to influence the sensitivity of detection. Similarly, an enzyme can be coated on the detector so that response is proportional to the rate of enzyme activity (Winquist et al., 1982). This has been demonstrated using the enzyme penicillinase (Caras and Janata, 1980). The enzyme was deposited on the surface of a H^+-sensitive FET as a cross-linked membrane with albumin. The small size of the detector, coupled with the small membrane thickness, means that the amount of enzyme required is low and the response times are short (1×10^{-4} i.u. and 25 seconds respectively). Although response times are short, mass-transfer and reaction considerations still apply. The major restriction on the use of ENFETs is the stability of the enzyme. Although the devices would be relatively cheap to mass produce, their limited stability may still restrict the cost effectiveness.

Direct enzyme–electrode interactions

Direct enzyme–electrode interactions stem from the concept of a fuel cell. In this system the enzymic oxidation of a compound is used to generate a flow of electrons

Enzyme-based sensors

through an external circuit. A simple example is the oxidation of methanol to formaldehyde using methanol dehydrogenase (Plotkin et al., 1981). The stoichiometry of the reaction can be written as

$$CH_3OH + \tfrac{1}{2}O_2 \longrightarrow CH_2O + H_2O$$
$$\text{methanol} \qquad\qquad \text{formaldehyde}$$

This can be written as two half-cell reactions

$$CH_3OH \longrightarrow CH_2O + 2H^+ + 2e^-$$

$$\tfrac{1}{2}O_2 + 2H^+ + 2e^- \longrightarrow H_2O$$

The fuel cell arrangement (Fig. 8.8) has an ion-exchange membrane being used to separate the two cells. In practice, a direct transfer of electrons from enzyme to electrode has yet to be achieved and so a mediator is required. This mediator is usually a redox dye which can be reduced by the enzyme and oxidized at the electrode; a commonly used mediator is the dye phenazine ethanosulphate (PES). Protons generated by the reaction pass through the cation-exchange membrane while the electrons generated flow through the external circuit resulting in a measurable current. The oxidation of methanol to either formaldehyde or formate has been demonstrated in a fuel cell device which, although generating only small currents, could form the basis of an extremely sensitive assay system. Although these fuel cells are not expected to offer a realistic opportunity for power generation in the near future, they may offer some solutions to specific problems (for example, the *in vivo* powering of heart pacemakers) and also some military applications, particularly the powering of communication devices.

For a purely analytical device the fuel cell construction is not essential. When only low currents are being drawn, separation of the two electrodes by an ion-exchange membrane is not necessary; hence the device can be very small. It is

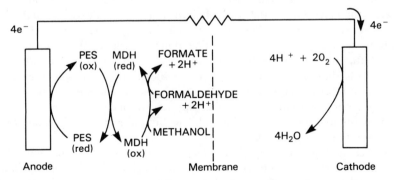

Fig. 8.8 Example of a simple bioelectrochemical fuel cell based on methanol dehydrogenase (MDH).
(From Plotkin et al., 1981)

envisaged that probes the size of a syringe needle could be used by diabetics for *in vivo* blood glucose measurement.

A distinction must be made between an enzyme electrode, which is the term used for an enzyme linked to an ion-selective electrode, and an enzyme-modified electrode, which can be regarded as a direct sensor with electrons being passed from the immobilized enzyme to a co-immobilized mediator and onto the electrode support matrix.

In the case of enzyme-modified electrodes, the current density required is much less than from a fuel cell, thus dramatically reducing the amount of enzyme required (Turner *et al.*, 1984). As the electrode acts as the electron acceptor in oxidase reactions, enzyme-modified electrodes have the important advantage of being insensitive to variations in the oxygen concentration of the sample to be measured. An additional advantage is that glucose-oxidase-modified electrodes using a ferrocene mediator have been shown to be linear over a wider range of concentrations than would be expected from the K_m value of the enzyme. As the response time of the probe is fast ($<$ 30 seconds), it is believed that this results from a modification of the enzyme's kinetic properties rather than mass transfer resistance. This glucose detector formed the basis of a patent application in 1981 and it is considered that sensors based on this technology will be ideal for glucose monitoring in diabetes. As our understanding of these 'enzyme-promoted' electron transfer reactions improves, the range of applications can be expected to increase widely.

Future developments are expected to allow the direct coupling of the enzyme to the electrode via a conducting linkage, thus removing the requirement for a mediator compound. This concept is fundamental for the development of 'biochips' (Weaver and Burns, 1981; Yanchinsky, 1982).

Other sensor devices

The measurement of human serum albumin (HSA) has been demonstrated using an 'affinity electrode'. Two identical titanium dioxide electrodes are constructed and one is then modified by the addition of the reactive textile dye Cibacron Blue F3G–1A (Lowe *et al.*, 1983). This dye has a high affinity for HSA and can be used in affinity separations (Chapter 5). The potential observed between a dye-modified titanium dioxide electrode and an unmodified reference electrode, when the system is illuminated, shows a linear response up to a concentration of 10 g ml^{-1} of HSA as the protein masks the photocell reaction. As the dye is a relatively stable molecule, the probe can be washed in 8M urea between measurements to allow regeneration. This technique may be applicable to the measurement of enzyme reactions.

Another approach to biological detection is the opto-electronic sensor: the change in concentration of one component causes a change in the colour of a dye. By placing the reaction chamber between a light emitting diode and a photocell, the concentration can be determined from the absorbance change. This approach has been demonstrated with the colour change associated with dye binding to

Enzyme-based sensors

protein and also with the enzymic detection of penicillin, glucose and urea, using pH-sensitive dyes. These can be regarded as a miniaturization of spectrophotometric techniques which take advantage of recent developments in microelectronics. Some examples of opto-electronic sensors are shown in Table 8.3.

Determination of biological oxygen demand

The sensor devices previously discussed have all been directed at the determination of a single compound. However, the determination of biological oxygen demand (BOD) is important in water pollution control, which requires that the contribution of all biodegradable molecules be assessed. Traditional methods require a long incubation period (5 days) at 20 °C and, because of a lack of viable alternatives, this has remained the method of choice since 1936. A recent alternative has been proposed using immobilized cells and an oxygen probe. This can be incorporated into an autoanalyser and results can be correlated with the conventional 5 day test. Although the response is slow (two samples per hour), it still represents a considerable time saving over the original method.

The search for cheaper, more stable, sensors has led workers to study the potential of probes using whole tissue preparations. Although there is a potential contamination problem, urea sensors based on ammonium probes and immobilized Jack bean meal (*Canavalia ensiformis*) have been shown to be more stable and at least as sensitive as sensors using purified urease.

Table 8.3 Performance characteristics of some opto-electronic sensors

Substrate	Response range (mM)	Output voltage change at 10 mM ($\Delta mV\ min^{-1}$)	Stability
Albumin	0.07–0.5 (5–35 mg ml^{-1})	—	»1 year
Penicillin G	0.3–5.0	45.5	>1 year
Ampicillin	0–10	24.1	
Cephaloridine	0–10	13.0	
Methicillin	—	0	
Vancomycin	—	0	
Cloxacillin	—	0	
Urea	0–40	125.0	$t_{\frac{1}{2}} \sim 17$ days
D-Glucose	0–70	1.5	$t_{\frac{1}{2}} \sim 7$–8 days

(From Lowe *et al.*, 1984)

Conclusions

It is apparent that there are a number of approaches which can be used to develop enzyme-based sensors (Guibault, 1984; Ichinose, 1986). In some cases their properties make them suitable for specific applications but it is clear that, in many applications, alternative sensors could be used. However, the approaches to biosensor development discussed here are not necessarily conflicting. While the potential ease with which bioelectrochemical sensors can be integrated with monitoring circuitry makes them very attractive, they are not suitable for all reagents. In general, the suitability of a given method must be assessed in the light of the application envisaged, and so developments in all areas can still be expected to lead to potentially commercial products.

Chapter 9

Approaches to enzyme modification

Introduction

It is normally considered that the properties and specificity of enzymes have evolved to the advantage of the host organism. However, although enzymes derived from a range of organisms have been useful in commercial or other applications, it may well be that certain properties of these proteins can be improved. For example, it may be advantageous to increase the heat stability or alter the pH optimum of an enzyme and thus improve the efficiency of a defined process. Alternatively, it may be desirable to enhance the activity of the enzyme towards one particular substrate or potential substrate.

Numerous methods have been developed for the modification of enzymes ranging from the very simple, e.g. exchange of cofactor, to highly sophisticated techniques involving molecular biology and computer-generated models. It is the aim in this chapter to describe some of the different approaches that have been taken to 'improve' enzyme performance and to evaluate the potential of these methods.

Selection of the appropriate source of enzyme

Initially, it is well worth considering whether the most suitable source of an enzyme is being used before embarking on a modification procedure. Enzymes catalysing the same reaction can often be found in many different organisms; for example, α-amylases can be isolated from animals, plants or micro-organisms and each has different properties, particularly with regard to thermostability.

The thermostability of α-amylases is of relevance in the enzymic hydrolysis of starch. Liquefaction is the disruption of the insoluble starch granule by heating (60 °C to 105 °C depending on the type of raw material), followed by partial hydrolysis to reduce the viscosity of the resultant solution (see Chapter 7 for details). With the less stable enzymes, the heat disruption and the partial hydrolysis have had to be separate stages within the overall process. Clearly, with the advent of commercially available thermostable enzymes (Table 9.1), particularly that from *Bacillus licheniformis*, the two steps may be combined into a single procedure. Thus, the simplest expedient of selecting the most appropriate enzyme source can greatly simplify a particular operational problem. This results in a more efficient process with a consequent saving in operational costs. Therefore, in some cases it may be possible to improve significantly a process without the need to modify enzymes at all. On the other hand, selection of the most appropriate source of an enzyme may just be prerequisite to modification procedures.

Substitution of bound metal ions

Many enzymes contain metal ions that are essential for activity. Whereas removal of the metal ion results in total loss of activity, it is often possible to substitute another cation with interesting results. A particularly illuminating example of the phenomenon is shown by certain of the D-glucose isomerases (see Chapter 7 for a more detailed discussion of the use of these enzymes).

Most of the enzymes that are known as D-glucose isomerases should more correctly be called D-xylose isomerases. This is because the primary role of these enzymes *in vivo* is the isomerization of D-xylose to D-xylulose. They are often inducible by D-xylose whereas D-glucose is ineffective. However, for the sake of

Table 9.1 Thermostability of α-amylases

Source of enzyme	Normal temperature of use	Maximum temperature of use
Porcine pancreas	40–45 °C	50 °C
Aspergillus sp.	55–60 °C	65 °C
Bacillus amyloliquefaciens	70 °C	85–90 °C
Bacillus licheniformis	92 °C	110 °C

The quoted temperatures should be regarded as approximate as the stability of the enzymes is dependent on the concentrations of substrate and Ca^{2+} ions.

Approaches to enzyme modification

convenience and because the reaction of industrial importance is the isomerization of D-glucose to D-fructose, the name D-glucose isomerase will be used here.

The glucose isomerase from the bacterium *Bacillus coagulans* strain HN-88 shows interesting properties in the presence of different metal ions that are applicable to the majority of this group of enzymes. In the absence of added metal ions the purified enzyme shows a substrate specificity for D-xylose with no activity towards either D-glucose or D-ribose (Table 9.2). Addition of 10mM $MnCl_2$ enhances the absolute activity of the enzyme and although D-xylose is still the substrate of preference, some isomerization of D-glucose and D-ribose is apparent. If 10mM $CoCl_2$ is used instead of the Mn^{2+} salt, then the substrate specificity of the enzyme is totally altered. D-Glucose and D-ribose are now the substrates of preference and activity towards D-xylose is relatively less. The addition of the metal-ion chelating agent, EDTA, totally abolishes activity towards all three potential substrates.

It may be suspected from the results obtained by addition of different metal ions to D-glucose isomerase that a different active site might be involved for each of the three sugars. However, it has been demonstrated by competition experiments that all three sugars utilize a common active site. Furthermore, the metal ion was not involved directly in the catalytic process and the binding of the cation occurs away from the active site. The effect of the cation is to cause a conformational change in protein structure that results in an altered binding of substrate. Thus, it is apparent that even relatively simple techniques such as substitution of the metal cofactor can have profound effects on the substrate specificity of an enzyme.

Table 9.2 Substrate specificity of D-xylose isomerase from *Bacillus coagulans*

	Relative activity[a]		
Additive (10^{-2} M)	D-glucose	D-xylose	D-ribose
None	0	4	0
$CoCl_2$	100	27	100
$MnCl_2$	16	100	20
$MgCl_2$	15	40	18
EDTA	0	0	0

[a] Relative activity is expressed as a percentage of that obtained using D-glucose as the substrate in the presence of 10^{-2} M $CoCl_2$.

(From Danno, 1970)

Covalent modification of enzymes

The chemical modification of proteins has been used widely as a tool for elucidating the mechanism of enzyme action. However, chemical modification may also be used to change the physical properties, substrate specificity or even the type of reaction catalysed by a particular enzyme, and it is some of these aspects that will be expanded upon. The immobilization of enzymes, by whatever method, should also be considered as a means of altering certain enzyme properties but, as this aspect of protein modification has already been dealt with (Chapter 6), it will not be reconsidered here.

Chemical modification of specific amino acids

The principle of the method is that covalent modification of a specific amino acid will alter the binding or catalytic properties of an enzyme (Kaiser *et al.*, 1985). Successful interpretation of these experiments requires an intimate knowledge of the structure of the protein under study, and for this reason much of the work has centred on the well-characterized proteinases. Usually, attempts have been made to alter either the pH optimum or substrate specificity of an enzyme and examples of both of these types of modification will be described.

Papain is a cysteine proteinase that requires a reduced thiol group at the active site and functional tryptophan residues. Modification of papain with hydroxyethyldisulphide to protect the sulphydryl group, followed by reaction with N-bromosuccinimide, resulted, in the oxidation of one or two of the five available tryptophan residues (either Trp-67 or both Trp-67 and Trp-177). Modification of Trp-67 alone resulted in a modified enzyme that had the same value of pK_a for the pH dependency of both k_{cat} and k_{cat}/K_m as the native papain. However, oxidation of both Trp-67 and Trp-177 resulted in an increase of approximately one unit for the pK_a values of both k_{cat} and k_{cat}/K_m. It has been postulated that modification of Trp-177 may result in a change in the hydrophobicity of the active site of papain with resultant pH effects on the kinetic constants.

It is possible by selective modification of amino acids to alter the relative catalytic activity of a proteinase towards protein and small molecular weight ester substrates. For example, modification of the serine proteinase subtilisin Carlsberg with tetranitromethane causes nitration of a single amino acid, Tyr-104. This resulted in a six-fold increase in proteolytic activity towards positively charged macromolecular substrates (e.g. clupein) but no increase in the degradation of a neutral protein (casein). However, surprisingly, no difference was observed in the relative rates of hydrolysis of the esters p-toluene sulphonyl-L-arginine methyl ester (positively charged) and benzoyl tyrosine ethyl ester (neutral substrate). The explanation for these apparently conflicting results is based on the concept of substrate size. Small molecules interact only with the primary binding site of subtilisin which is unaffected by the modification procedure. In contrast, the larger protein substrates also interact with a secondary binding site that contains the nitrated Tyr-104. The nitration of subtilisin results in a lowering of the pK_a of the hydroxyl group on Tyr-104, effectively introducing a negative charge into the

Approaches to enzyme modification

secondary binding site. Therefore, positively charged macromolecules are bound more readily by the derivatized enzyme whereas the small molecular weight ester substrates are unaffected.

Although the chemical modification of specific amino acids can produce profound effects on the activity of enzymes, the technique is rather limited. Most modification reagents are not absolutely specific for a certain amino acid and it is often difficult to limit the extent of the reaction. Consequently, this method has not really found widespread application in an industrial context, although it does have the advantages of simplicity and cheapness.

Enzymic modification of enzymes

In principle it is possible to alter the properties of an enzyme by treatment with proteinases or glycanohydrolases. The use of various glycanohydrolases to modify the oligosaccharides of glycoproteins has already been considered in the context of medical applications (Chapter 5). Limited proteolysis can also be used to modify the activity of an enzyme and it is this topic that will be discussed here.

Proteolytic modification occurs *in vivo* and is often a prerequisite to the formation of an active enzyme (e.g. the proteolysis of trypsinogen to produce trypsin). There is one excellent example of the use of a proteinase to modify an enzyme for commercial use. The DNA-directed DNA polymerase I of *Escherichia coli* can synthesize DNA under certain closely controlled conditions. Paradoxically, the same enzyme is also able to degrade DNA as it possess both $5' \rightarrow 3'$ and $3' \rightarrow 5'$ exonuclease activities. Although these apparently conflicting properties can be reconciled in terms of the known function of DNA polymerase *in vivo*, the exonuclease activity (particularly the $5' \rightarrow 3'$ exonuclease) is a considerable disadvantage during the synthesis of DNA *in vitro* (see later in this chapter). Consequently, modification of DNA polymerase such that the $5' \rightarrow 3'$ exonuclease is no longer present would be of considerable value.

The removal of the $5' \rightarrow 3'$ exonuclease activity may be achieved simply by the limited proteolysis of DNA polymerase (Jacobsen *et al.*, 1974). Treatment of the enzyme with subtilisin Carlsberg results in the generation of two fragments which can be separated by chromatography on hydroxyapatite. The larger fragment, known as the Klenow enzyme, possesses the polymerase and the $3' \rightarrow 5'$ exonuclease activities, whereas the smaller N-terminal fragment has only the $5' \rightarrow 3'$ exonuclease (Table 9.3). Thus a simple proteolytic treatment of DNA polymerase results in an enzyme that has lost one of its three activities. The Klenow enzyme is sold commercially for use in DNA sequencing and oligonucleotide mutagenesis procedures (see later).

It is interesting to see how manufacturing techniques have advanced. Several companies are now offering a preparation of Klenow enzyme that has been produced by gene-cloning technology, thus obviating the need for the proteolytic step. However, because of patent protection, the majority of companies still produce Klenow in the traditional way.

Table 9.3 Activities of DNA polymerase fragments generated by limited proteolysis

	Activity (mol of deoxynucleotide incorporated or released min^{-1} mol $enzyme^{-1}$)		
Fragment	Polymerase	$5' \rightarrow 3'$ exonuclease	$3' \rightarrow 5'$ exonuclease
Large	227.0	1.1	40.0
Small	<0.2	135.0	0.6

(From Jacobsen et al., 1974)

Enzyme–coenzyme complexes

The present commercial exploitation of enzymes has been restricted largely to the simple hydrolases whereas the benefits of utilizing dehydrogenases and kinases for synthetic reactions are still unrealized. In part this is because the use of coenzymes such as ATP and NAD^+ is prohibitively expensive unless some method of reclaiming and recycling is available. A number of techniques have been developed for the retention of coenzymes within enzyme-reactor systems by immobilization onto soluble or insoluble matrices. However, one of the more exciting ideas is the possibility of covalent attachment of the coenzyme to the active site of an enzyme.

Several stable enzyme–coenzyme complexes with endogenous catalytic activity have been developed. Derivatives of NAD^+ have been coupled to lactate dehydrogenase by several methods and the resultant complexes have intrinsic activities close to the theoretical values estimated from the stoichiometry of coenzyme binding (Gacesa and Venn, 1979). However, the use of these complexes will probably be limited by the low concentrations of NAD^+ that can be bound to the enzyme and by the need to recycle the reduced coenzyme.

The concept of covalent attachment of a cofactor may be extended to enzymes other than the oxidoreductases and kinases. For instance, there is the possibility of attachment of a redox coenzyme analogue at the active site of a simple hydrolytic enzyme, thus creating a totally new activity (Kaiser and Lawrence, 1984). There are a number of points to consider and some of the principal investigators in this field have drawn up five rules to increase the chances of success:

(1) The enzyme should be available highly purified and in quantity.
(2) The X-ray structure of the enzyme should be known.
(3) There should be a suitably reactive amino acid sidechain close to the active site.
(4) The attachment of the cofactor should cause a profound change in enzyme activity.
(5) The access of potential substrates to the active site should not be hindered.

Approaches to enzyme modification

The most significant work in this area has been the development of the semi-synthetic flavo-papains. They have been derived by the covalent attachment of flavin analogues to the active-site cysteine. (Cys-25) of the proteolytic enzyme papain. As Cys-25 is essential for the normal hydrolytic activity of papain, substitution at this position results in the inability to hydrolyse proteins. This is useful as the loss of proteolytic activity can be used to monitor the progress of substitution with cofactor, and also any new activity that may be generated will be more readily observable.

Essentially, two types of flavins, substituted at various positions with either a bromoacetyl or a bromomethyl group (Fig. 9.1), have been used to derivatize the free sulphydryl group of papain. The flavo-papains derived from the bromoacetyl-substituted isoalloxazines are more effective catalysts than those produced from the bromomethyl-containing compounds. The reason for this is that the carbonyl group in the former compound is able to hydrogen bond to both the backbone amide of Cys-25 and the sidechain amide group of Glu-19 (Fig. 9.2). This hydrogen bonding ensures that the substituted isoalloxazine ring system is held flexibly in close juxtaposition to the hydrophobic substrate binding site.

The flavo-papain derived from 8α-bromoacetyl-10-methylisoalloxazine shows many of the required characteristics of an oxidoreductase. The semi-synthetic enzyme oxidizes dihydronicotinamide derivatives with a second-order rate enhancement of up to 500-fold compared with the corresponding chemical reaction, and also displays saturation kinetics. Furthermore, the second-order rate constant for the semi-synthetic enzyme is comparable with those obtained for native flavo-enzymes (Table 9.4). A further feature of the flavo-papains that has been demonstrated with the 7α-bromoacetyl-10-isoalloxazine-substituted enzyme is the preferential abstraction of the 4A (proR) hydrogen atom of NADH (Fig. 9.3)

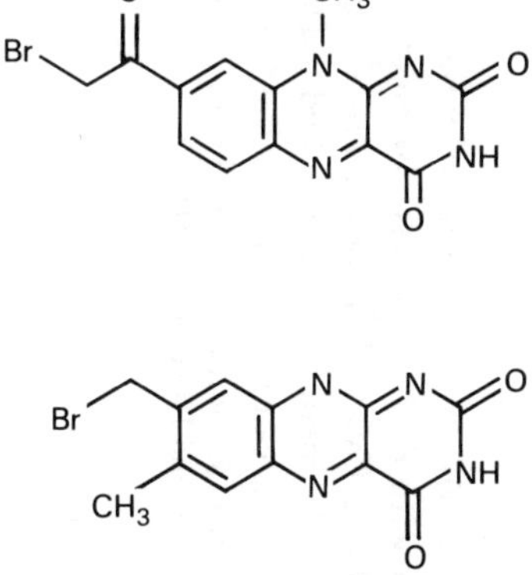

8α-Bromoacetyl-10-methylisoalloxazine

8α,h bromomethyl-7-methylisoalloxazine

Fig. 9.1 Structures of some flavin analogues. The sulphydryl group of papain (Cys-25) may be alkylated with either of these compounds to produce the corresponding enzyme–flavin derivatives.

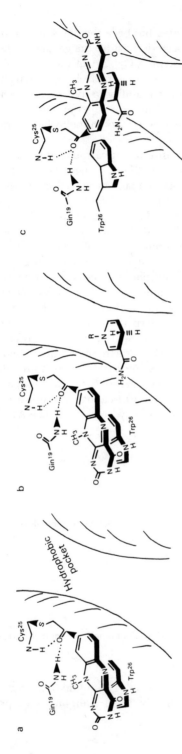

Fig. 9.2. Active site of a flavo-papain derivative. (a) Active site of the semi-synthetic enzyme, derived by the alkylation of papain with 8α-bromoacetyl-10-methylisoalloxazine. The acetyl sidechain of the flavin moiety is hydrogen bonded to the Gln-19 and Cys-25 backbone. The flavin is participating in a charge-transfer complex with Trp-26. (b) Michaelis complex. The dihydronicotinamide is embedded within the hydrophobic groove of the flavo-enzyme. (c) ES′ intermediate. The flavin-Trp-26 charge-transfer complex has been disrupted and the flavin now lies directly over the nicotinamide substrate. The pro-R hydrogen is shown as the species being transferred to the N-5 position of the flavin. (From Kaiser and Lawrence, 1984)

Approaches to enzyme modification 129

Table 9.4 Second-order rate constants for flavo-enzymes. The data have been obtained for the oxidation of N′-hexyl-1-4-dihydronicotinamide under comparable conditions

Enzyme	k_{cat}/K_m ($\text{M}^{-1}\text{sec}^{-1}$)
NADH-specific FMN oxidoreductase	3.3×10^5
NADPH-specific FMN oxidoreductase	8.5×10^5
Old Yellow enzyme	6.1×10^2
Flavo-papain	5.7×10^5

(From Kaiser and Lawrence, 1984)

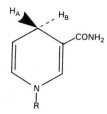

Fig. 9.3 The structure of the dihydronicotinamide ring.

during oxidation. The stereospecificity of hydrogen atom abstraction at C4 is a characteristic of native oxidoreductases.

The flavo-papains clearly demonstrate the feasibility of totally changing the reaction catalysed by an enzyme. In particular, the ability to utilize a cheap and readily obtained protein such as papain to generate catalytic activities normally associated with more expensive enzymes has great potential. Although the flavo-papains represent the most widely studied example of semi-synthetic enzymes, other model systems have been studied. Clearly this is an area of research that is ripe for development.

Non-specific mutagenesis

Various methods of *in vivo* mutagenesis have been used to produce modified or new enzyme activities.

Microbial evolution
One method of designing new enzyme activities is to select specific mutants of

bacteria by growth of the organism on poorly or non-metabolized substrates. This technique is called microbial evolution (Clarke, 1980). In essence, a proportion of a population of bacteria is often able to adapt to a new growth compound. In some cases this involves the production of a mutated form of an enzyme whereas, in other circumstances, the altered phenotype results from a change in gene regulation. Several well-characterized examples of microbial evolution have been documented.

The bacterium *Klebsiella aerogenes* grows only slowly in the presence of the pentose D-arabinose. However, it is possible to select mutants with an enhanced growth rate. In these mutants L-fucose isomerase, which has minimal activity towards D-arabinose, has become deregulated and the enzyme is produced constitutively. Consequently, the new constitutive enzyme, which is produced in greatly increased quantities, is able to cope, albeit inefficiently, with the isomerization of D-arabinose even though no structural gene mutation has occurred. Subsequently, it is possible to isolate mutants that have a change in the structural gene such that the L-fucose isomerase has an increased affinity for D-arabinose. Therefore, because of the rapid growth and division of micro-organisms, it is possible to generate new enzyme activities in a relatively short time.

Another example of microbial evolution is the elegant work on the amidase of *Pseudomonas aeruginosa*. Again, the evolutionary process is characterized by the appearance of regulatory and then structural gene mutations. Whereas the amidase of the wild type *P. aeruginosa* has significant activity towards acetamide and propionamide, various mutants were derived in which the enzyme could hydrolyse such compounds as valeramide and phenylacetamide and in some cases had lost the ability to utilize the original substrates.

Although the technique of selecting mutants is really applicable only to micro-organisms, it does offer a way of obtaining enzymes that are active against defined substrates with little or no requirement to understand the enzyme mechanisms.

Site-specific mutagenesis

Although radiation (e.g. uv light) and chemicals (e.g. hydroxylamine) have been used to promote random mutations in DNA, it is now possible to use procedures that are much more specific (Smith, 1982). In conjunction with modern recombinant DNA technology, these techniques create mutations at pre-determined sites within a DNA molecule, thus causing predictable changes in the properties of the encoded enzyme.

Chemical mutagenesis
Certain chemicals can alter the structure of selected bases within the DNA molecule. One of the more common techniques is to deaminate cytosine to produce uracil by treatment of the DNA with sodium bisulphite (uracil is not normally present in DNA but it acts in an equivalent way to thymine). This change is mutagenic because, during the subsequent replication of DNA *in vivo*, cytosine would have been the template for guanine incorporation whereas uracil

Approaches to enzyme modification

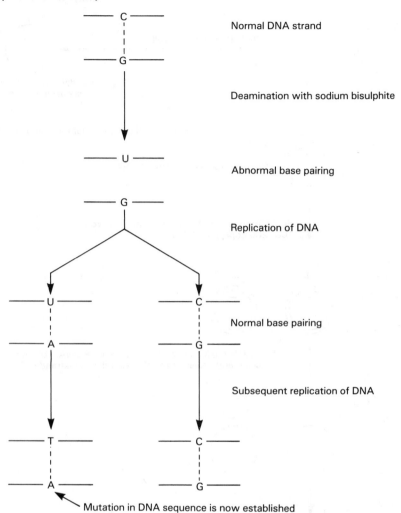

Fig. 9.4 Effect of cytosine deamination on DNA replication.

will base pair with adenine (Fig. 9.4). It is essential that these experiments are done using a strain of *E. coli* that is deficient in the DNA uracil N-glycosidase repair mechanism, otherwise the uracil will be edited out and replaced by the correct base.

The beauty of the sodium bisulphite method is that the deamination can be limited. Cytosine is deaminated efficiently by sodium bisulphite only when present in a single-stranded piece of DNA. Therefore, by the use of specific endo- and exonucleases to cut a small part of one strand of the normally double-stranded DNA, it is possible to pre-determine the site of mutagenesis. Alternatively, single-stranded DNA may be generated by the use of M13 phage vectors (see later in this

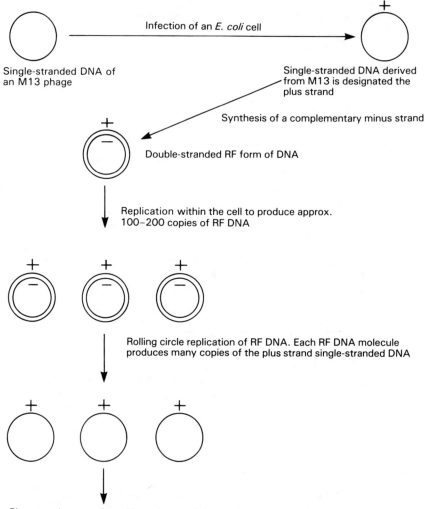

Fig. 9.5 Life cycle of the filamentous phage M13.

chapter). The main disadvantage of this method is that only a cytosine to uracil transition is possible. Although other chemicals are able to cause mutations in other bases, the method is still somewhat limited in its applications.

Oligonucleotide mutagenesis
One of the greatest advances in enzyme technology over the past few years has been the development of oligonucleotide mutagenesis. In principle, this method allows a single amino acid change to be made at a pre-determined site in a protein by specific mutation of the encoding gene. This technique has been facilitated

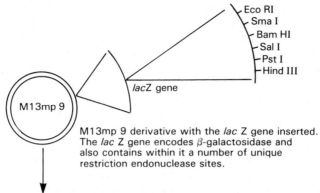

M13mp 9 derivative with the *lac* Z gene inserted. The *lac* Z gene encodes β-galactosidase and also contains within it a number of unique restriction endonuclease sites.

Cut both the DNA to be cloned and M13 mp 9 with the same restriction endonuclease. Use DNA ligase to join the foreign DNA insert and the M13 mp 9 vector.

Transform a β-galactosidase-deficient strain of *E. coli* with the ligation mixture and plate out onto agar containing isopropyl β-D-galactopyranoside (an inducer of β-galactosidase) and X-gal (a chromogenic substrate for β-galactosidase).

M13 mp 9 derivatives containing fragments of NA inserted into the *lac* Z gene will not express β-galactosidase (insertional inactivation) and will produce white plaques. Non-recombinant M13 mp 9 derivatives will produce blue plaques as a result of the expression of the intact *lac* Z gene.

Select white plaque, isolate phage DNA and sequence from the Hind III site to determine the orientation of insertion.

Fig. 9.6 Selection of M13 phage containing recombinant DNA.

greatly by the dramatic improvements in the chemical synthesis of oligonucleotides and in the characterization and use of filamentous phages, e.g. M13.

To appreciate the mechanism of oligonucleotide mutagenesis, it is necessary to know something about the life cycle of the M13 phage. The genetic information of M13 is enshrined in a single-stranded DNA molecule (ssDNA). When the phage infects susceptible *E. coli* cells (those that contain the F episome), the ssDNA is converted into the double-stranded replicative form (RF DNA). Subsequently, approximately 200 copies of the RF DNA are made within the *E. coli* cell before the final stages of phage maturation. The RF DNA molecules are used as templates for the production of multiple copies of the plus strand ssDNA that will be packaged into new viral particles (Fig. 9.5) and the intact M13 phages are then released without causing either the lysis or the death of the infected *E. coli*. Consequently, if M13 DNA is isolated from infected *E. coli*, then it will be in the double-stranded RF form whereas extraction from the culture supernatant will yield ssDNA.

The first step is oligonucleotide mutagenesis is to clone the gene of interest into the commercially available RF DNA of an M13 phage derivative. These M13 derivatives have already been genetically engineered to contain part of the gene encoding β-galactosidase (*lac* Z gene) (Gronenborn and Messing, 1978). The foreign DNA is ligated into the *lac* Z gene of the phage and the recombinant material introduced into a β-galactosidase-deficient strain of *E. coli* by transformation. Consequently, growth of the transformed *E. coli* on agar plates containing the chromogenic substrate for β-galactosidase (5-bromo-4-chloro-3-indolyl β-D-galactopyranoside (X-gal)) will allow resolution of recombinant (insertionally inactivated β-galactosidase gene) phage (Fig.9.6). The foreign DNA may be inserted randomly in either direction, but for oligonucleotide mutagenesis it is important that the coding gene is ligated into the plus strand of the phage (Fig. 9.6). The orientation of the inserted gene may be established readily by partial dideoxynucleotide sequencing of the DNA. Plus-strand DNA containing the coding sequence of the enzyme to be mutagenized may be isolated from the released phage that are present in the supernatants of selected bacterial clones. In principle there is no limit to the length of foreign DNA that can be inserted into the genome of the M13 phage.

Essentially two methods (either phosphotriester or phosphite–triester chemistry) are available for the solid-phase synthesis of the mutagenic oligonucleotide. Appropriate nucleotide derivatives and chemicals are readily available and the techniques of synthesis are well characterized and easy to use (Gait, 1984). The oligonucleotide should be at least fourteen nucleotides long (a 14mer) to minimize random association, and conventionally a 19mer is often found to be sufficiently selective without the need to synthesize very large molecules. Also, the mutation should be placed close to the centre of the oligonucleotide to minimize the possibility of mismatch-base replacement by either the Klenow enzyme (used in the next step) or the host bacterium. The sequence of bases in the mutagenic oligonucleotide should be complementary to the coding strand and will form part of the minus strand in the RF DNA. The oligonucleotide is annealed to the plus strand and the synthesis of the minus strand is completed *in vitro* using the Klenow enzyme and DNA ligase (Fig. 9.7).

Transformation of a strain of *E. coli* with the RF DNA will produce a mixed population of mutant and non-mutant phage, i.e. those derived from the minus and original plus strands, respectively (Fig. 9.5). In most strains of *E. coli* the mutants will constitute less than 5% of the total phage population because of the ability of the bacterium to correct the defect in base pairing. However, use of mismatch-repair deficient strains of *E. coli* can ensure that approximately 45% of the phage population will encode for the mutant form of the enzyme. Reinfection of a fresh lawn of bacteria with a low titre of the mixed population of phage will enable pure mutant clones to be selected (at low titre each plaque is assumed to be derived from a single phage). Those plaques containing mutant phage may be detected by using a radioactive DNA probe based on the mutant oligonucleotide

Approaches to enzyme modification 135

Isolate ssDNA from a recombinant phage that contains the gene that is to be mutated.

Add mismatch oligonucleotide. This will base pair (with the exception of the mutation) with the plus strand.

Add Klenow enzyme and DNA ligase to complete the synthesis of the minus strand *in vitro*

Transform a mismatch-repair deficient strain of *E. coli*

Production of multiple RF DNA; some will contain the mutation (up to approx. 45%) and others will not

Transfect fresh *E. coli* at low titre and screen individual plaques for the presence of mutants, either by using a labelled oligonucleotide probe or by sequencing from the Hind III site of M13 mp 9.

Select M13 mp 9 that contains that required mutation

Clone the gene that contains the mutation into a suitable host organism.

Expression of site-directed mutagenized protein.

Fig. 9.7 Oligonucleotide mutagenesis using M13 phage vectors.

or by sequencing the gene. The mutated gene may then be cloned into a suitable expression vector and the modified enzyme synthesized by an appropriate organism.

Although the techniques of oligonucleotide mutagenesis are relatively new, a number of important papers have already been published. For example, subtilisin BPN' from *Bacillus amyloliquefaciens* has been used as a model for several types of modification. Subtilisin is readily inactivated by oxidation of the methionine residue at position 222. Experiments, in which Met-222 has been replaced by each of the other nineteen amino acids, established that the susceptibility to oxidation could be abolished, although with partial loss of catalytic activity. However, one derivative of subtilisin (Met-222→Cys-222) had a much improved stability against oxidation but still had a k_{cat}/K_m that was 56% of the value of the wild type (Estell *et al.*, 1985).

Oligonucleotide mutagenesis of subtilisin BPN' has been used to demonstrate how the pK_a of an active site group may be modified (Thomas *et al.*, 1985). Mutation of Asp-99 to Ser-99 will result in the loss of a negative charge. It can be predicted from models that this will cause a lowering in the pK_a of the active site amino acid His-64 because of changes in electrostatic interactions within the enzyme. In practice, this mutation results in a reduction of the pK_a of His-64 from 7.17 to 6.88 at low ionic strength (0.1). At high ionic strength (1.0) this effect is abolished, and the mutant and wild-type enzymes have the same pK_a value for His-64. This is to be expected if the interaction between Asp-99 and His-64 is electrostatic in the wild-type enzyme. These experiments clearly show the potential of the method for adjusting the pH profiles of enzymes.

The techniques of oligonucleotide mutagenesis have developed extremely rapidly over the last few years and appropriate kits are now available commercially. This opens the door for enzymologists to modify enzymes specifically using this sophisticated technique without the need for highly specialized research teams of molecular biologists. Clearly, oligonucleotide mutagenesis, in conjuction with computer modelling (see Chapter 10), is fast becoming one of the most important new tools available to the enzymologist and will be a major growth area during this decade.

Chapter 10

Future prospects

Introduction

The past twenty-five years or so have seen a remarkable expansion in the enzyme industry. However, it has not all been without mishap, and one has only to recall the disaster of the late 1960s when enzyme-containing washing powders were considered to be harmful, with a resultant slump in the market. Predicting the growth of the enzyme market over the next twenty-five years and identifying future applications will be difficult, and it is perhaps not appropriate to attempt this in the context of this book. Decisions relating to the novel applications of enzymes will depend on the overall economics of a process, of which the scientific feasibility is but one factor in the equation. Therefore, in this chapter we have aimed to highlight what we consider to be some of the major challenges to the enzymologist over the next few years, each of which, when solved, should result in significant advances in the underpinning science.

Throughout the book we have tried to indicate those areas of enzyme technology that we felt would be significant growth points. For example, the application of gene-cloning techniques (Chapter 2) and site-directed mutagenesis (Chapter 9) is clearly going to have profound effects on the subject in the future. In this chapter we have chosen to discuss four research areas: prediction of enzyme folding/structure, the use of enzymes in organic solvents, artificial enzymes, and coenzyme regeneration. Each may be identified as a fundamental problem in which advances in our basic knowledge will foster the development of significant new applications. This selection of four topics is intended to illustrate the future potential of enzyme engineering and is in no way exclusive of other possibilities. Also, each of these topics covers areas in which promising advances have already been made and in which a way ahead is at least apparent.

Prediction of enzyme folding/structure

Experiments conducted more than twenty-five years ago established that the primary sequence of a polypeptide chain contained all of the necessary information for the correct folding of a protein. However, if provided with an amino-acid sequence for a novel polypeptide chain, biochemists are still not able to predict accurately the pattern of folding that will occur. It has become increasingly important to be able to determine the three-dimensional structure of an enzyme, particularly if the technique of site-directed mutagenesis is to be used to modify properties. Clearly one needs to know the position and type of change in amino-acid structure that should be made for effective results. In other words, there has to be a mechanistic rationale if changes are to be made in anything other than a random manner.

To be able to determine the three-dimensional structure of an enzyme it is necessary to know the primary sequence. Methods for obtaining polypeptide primary structure have advanced enormously (Hunkapiller and Hood, 1983). Proteins may be sequenced directly using variations on the Edman degradation procedure (Fig. 10.1); in practice, with good quality reagents and equipment, it is possible to sequence from thirty to seventy amino-acid residues in a single run before losses and side reactions become prohibitive. Both manual and automated methods are available. Nowadays, it is often preferable to determine the primary structure of a protein indirectly by sequencing the gene (Sanger *et al.*, 1977). Isolation and cloning of the appropriate piece of DNA, followed by sequencing using the M13 phage system with Sanger's dideoxy chain termination method (Fig. 10.2), is often quicker and more accurate than working directly on the protein. In practice, the two methods are complementary as the DNA approach will not reveal any post-translational modifications that may have occurred.

Accurately predicting the three-dimensional structure of an enzyme from the amino-acid sequence alone is still not possible despite the development of algorithms and the advent of supercomputers (van Brunt, 1986). This inability to predict three-dimensional structure accurately is the result of two factors. The first is that most of the algorithms assume that all the amino acids make some significant contribution to the overall structure. While this is true, in practice it is undoubtedly the subtleties of interaction that have a profound effect on the protein conformation. For instance, there are many cases where a single amino-acid substitution has not affected the structure, stability or function of an enzyme. On the other hand, there are examples where a single alteration of a critical amino acid causes profound changes in properties. The problem is trying to identify which amino acids are critical, understand why, and to try to produce a model that is generally applicable to all proteins. Studies on the genetic analysis of protein-folding pathways and the use of enzyme derivatives produced by site-directed mutagenesis should help to resolve the problem and there are some rational explanations beginning to emerge (King, 1986). The second problem in trying to predict protein structure is that calculating the position of individual amino acids to within even a few angstroms takes a large amount of computing power. Unfortunately, it is often the sub-angstrom distances and interactions that

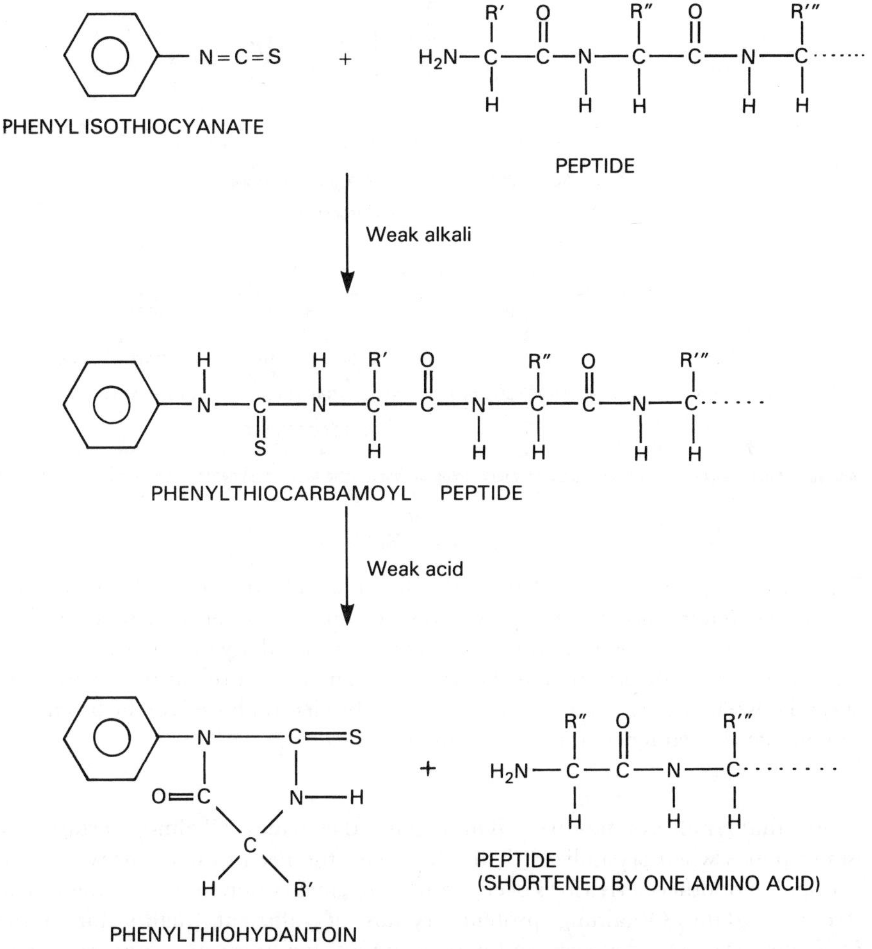

Future prospects

Fig. 10.1 Sequencing of polypeptides using the Edman reagent.

are critical to the functioning and stability of an enzyme. Nevertheless, computer modelling and molecular graphics have come a long way and, with the ever-increasing power of computers and improvements in the algorithms, it is likely that some day predictions of this sort will be accurate, reliable and commonplace.

Current methods for determining protein three-dimensional structure are limited to X-ray diffraction analysis (Ulmer, 1983). Conventionally, crystals are bombarded with X-rays and the diffraction patterns are analysed. The limitation of conventional X-ray diffraction is that, although the positions of the larger atoms may be determined with precision, the precise location of the protons is less clear. More recently, use has been made of synchrotron X-ray sources which have enabled data to be obtained faster. Also, because the wavelength is tunable it has the potential to eliminate the need to obtain isomorphous crystals. It is interesting

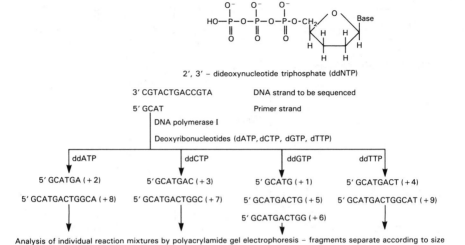

Fig. 10.2 Sequencing of DNA using the dideoxynucleotide method. The primer strand is extended using DNA polymerase I and the sequence of bases will be dependent on the sequence in the complementary strand. By incorporating one of the dideoxynucleotides into each of four reaction mixtures premature termination of synthesis will occur. Subsequent size analysis by electrophoresis of the fragments will enable the sequence to be determined.

to note that families of site-directed mutagenized enzymes will almost certainly be isomorphous when crystallized. This eliminates the need to make heavy-metal-containing or other derivatives of crystalline enzymes to obtain three-dimensional structural data. Obtaining protein crystals of sufficient quality for X-ray diffraction studies is not easy and may be impossible in certain cases. However, currently the structures of some two hundred proteins have been determined to atomic resolution.

One problem with X-ray diffraction is that the protein must be crystalline and this really begs the question of what the structure of the molecule might be in solution. A technique that has shown great potential is two-dimensional proton nuclear magnetic resonance (see references cited by Ulmer, 1983). With the advent of superconducting magnets producing high-strength magnetic fields, in conjunction with two-dimensional nuclear Overhauser spectroscopy, it is already possible to make individual assignments for almost all of the resonance lines in a ^{1}H-nmr protein spectrum. Furthermore, semi-quantitative information on proton–proton distances can be obtained at the 2 to 5 angstrom level.

The analysis of protein structure by nmr is based on the fact that the resonance value for an individual proton will depend on its immediate environment. In essence the method correlates data from two types of two-dimensional ^{1}H-nmr analysis. First, interactions of protons that are close in terms of covalent linkage, i.e. interconnected by a distance of two or three covalent bonds, are determined (known as through-bond scalar J-connectivities). These data are then compared with interactions of protons that are located short distances from each other in spatial structure (known as through-space dipolar NOE-connectivities). By correlating these sets of data, it is possible to obtain a three-dimensional spatial proton map of the protein. Most of the structures that have been determined so far have been related to known X-ray diffraction data. However, with suitable refinement this method has the potential to provide information about protein structure at sub-angstrom resolution and without the need for crystallization. The elucidation of protein three-dimensional structure is still one of the most time-consuming aspects of enzyme technology and any advances in this area are likely to make a considerable impact.

Use of enzymes in organic solvents

Many conventional industrial processes are based on systems that utilize organic solvents and hydrophobic compounds. However, in the past it has been considered that enzymes do not tolerate non-aqueous solutions and therefore they would be unsuitable as catalysts in these circumstances. These ideas were based on observations made long ago; for example, miscible organic solvents such as acetone have been used for the selective precipitation of proteins during purification procedures. Also, in those cases where enzymes were found to work in non-aqueous systems, kinetic parameters were often adversely affected. A typical observation is that noted with the trypsin-catalysed hydrolysis of benzoyl arginine ethyl ester in dioxane–water mixtures. In this case the K_m value increases 1000- to 5000-fold in 80% (v/v) dioxane compared with a completely aqueous environment. This change in K_m was attributed to an increase in the long-range repulsion between positively charged trypsin and substrate molecules, which increases with decreasing dielectric constant.

A number of enzyme-catalysed transformations in organic solvents have been investigated, particularly those reactions that involve substrate(s) and/or product(s) that are particularly water-insoluble (Klibanov, 1986). It is important to differentiate between enzyme performance in water-miscible and water-immiscible solvents. Experiments established early on that, once the concentration of an organic, water-miscible solvent exceeded about 50% (v/v), then all but the toughest enzymes would be inactivated (there are some important exceptions to this rule — see later on in this chapter). Unfortunately, this concentration is often insufficient to solubilize effectively a non-water soluble substrate or product. This observation led to the false premise that a water-immiscible solvent would be even more likely to denature an enzyme. However, this has not proved to be so and the potential of using water-immiscible solvents is much greater than was initially realized. It is these water-immiscible systems that will be considered in greater detail here.

The proportion of water that is present in the overall system is an important consideration. If the water content is high enough, then a water-in-oil micro-

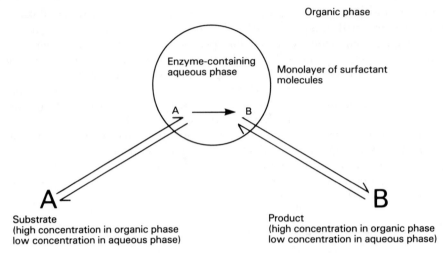

Fig. 10.3 Enzyme-mediated catalysis in reverse micelles.

emulsion is formed with the enzyme contained in the aqueous phase. However, if the water content is reduced, then, in the presence of a suitable surfactant, reverse micelles will be formed (Fig. 10.3). In practice, it is difficult to decide between what is a reverse micelle and what is a water-in-oil emulsion, and the term reverse micelle is often used to describe both situations (Luisi and Laane, 1986). When the water content is even lower, then the enzyme forms insoluble particles in an organic solvent and the result is a heterogeneous catalysis system.

All three of the above-mentioned variations have been used and each has advantages in a particular case. For example, the conversion of cholesterol to cholestenone has been studied in both reverse micelles and in the ultra-low water system. Both the reaction rate and operational stability of the enzyme in reverse micelles (cetyltrimethylammonium bromide/octanol/octane) were higher than in aqueous media. However, better results were obtained when whole *Nocardia* cells (containing cholesterol oxidase) were used in dry organic solvents. As both cholesterol and cholestenone are more soluble in organic solvents, this means that the concentration of substrate may be increased 100-fold compared with a completely aqueous reaction mixture. The organic phase in a biphasic system acts as both a reservoir of cholesterol and as a sink for the cholestenone. In some cases, reverse micelles or some other biphasic system is a necessity for a satisfactory reaction to occur. The specific reduction of 20-keto steroids to their corresponding 20-β-hydroxysteroids requires the combination of 20-β-hydroxysteroid dehydrogenase and a NADH-regenerating system. By using reverse micelles or two-phase solvents, the steroids are soluble in the organic phase but are able to partition into the aqueous phase which contains the enzymes and coenzyme. Thus the overall reaction has been compartmentalized in a way that is not unlike living cells.

Enzymes can behave quite atypically in dry organic solvents and may acquire useful properties. Although a particular enzyme has an absolute requirement for some water so that it may adopt the appropriate active conformation, only a minimal amount is necessary in practice. In aqueous solution the water concentration is 55.5M and only a minute proportion of this is used to solvate the enzyme, although some of the rest may be a participating substrate in the reaction. The essential water is very tightly bound and will not be removed by conventional drying procedures. Therefore, a dried enzyme powder will always contain the minimum amount of water which is essential for activity. When the dried preparation is added to an organic solvent, it will not dissolve but will remain as discrete particles and the residual water will partition between solvent and enzyme. It is for this reason that it is essential that the solvent is water-immiscible, otherwise water will be removed and the enzyme will not retain activity. Additional advantages of using low-water systems are that enzymes may be more stable and may catalyse novel reactions.

The chemical and structural changes that cause irreversible thermal activation of enzymes all require water and therefore one might predict an increased stability in organic solvents. This appears to be the case and is beautifully illustrated by some experiments using porcine pancreatic lipase. If a dried powder of the lipase is placed in water at 100 °C, then inactivation is almost instantaneous. However, if instead the enzyme powder is placed in 2M-n-heptanol in tributyrin containing either 0.8% or 0.015% water, then the lipase has a half-life at 100 °C of 10 min or 12 hours respectively (Fig. 10.4). Furthermore, not only is the enzyme relatively stable at this temperature but it is capable of catalysing a transesterification reaction five times faster than at 20 °C.

The lack of water increases the degree of rigidity of the enzyme molecules and this can have a number of effects on specificity. Many of the enzymes that have been used in organic solvent systems are hydrolases. Clearly, if the concentration of water is low and another suitable nucleophile is available (e.g. alcohols, amines, thiols, etc.), then novel reactions will occur. For example, lipases from either porcine pancreas or yeast will, in the absence of water, catalyse esterification and

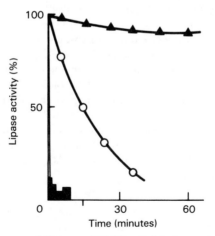

Fig. 10.4 The thermostability at 100 °C of dried porcine pancreatic lipase placed in 0·1M phosphate buffer, pH 8.0 (■), or in 2M-n- heptanol in tributyrin containing either 0·8% water (O) or 0·015% water (▲).
(From Zaks and Klibanov, 1984)

transesterification reactions. Thus, in the example cited above of the porcine pancreatic lipase working in n-heptanol in tributyrin, the major reaction is transesterification with production of the monoester of butyric acid and dibutyrin rather than simple hydrolysis. The enzyme was able to utilize a wide range of alcohol substrates for transesterification of tributyrin in this reaction scheme. Interestingly, if the alcohol contained a bulky substituent, e.g. *tert*-butanol, then the substrate specificity could be controlled by altering the concentration of water. At the lowest water content (0.015%) *tert*-butanol was not utilized because the enzyme was too inflexible to accommodate the bulky substituent within the active site, but at 0.7% water content sufficient flexibility was introduced to enable the reaction to proceed. The initial rates of reaction are lower in organic solvents compared with aqueous solutions (typically 20–50%) but this is unlikely to be a problem as the enzymes are relatively cheap and there is the potential of producing novel products.

Although most advances have been made using enzymes in water-immiscible solvents, there are examples of useful applications in miscible systems. For example, the optical resolution of organic acids and the synthesis of peptides using proteinases both use miscible solvents. A challenging new area of development has been the use of enzymes and miscible solvent systems for the degradation of lignin. Lignin is the largest source of renewable aromatic chemicals on Earth; however, the material is notoriously difficult to degrade selectively by either chemical or enzymic methods. Recently, a peroxidase-type enzyme (ligninase) has been isolated in small quantities from a fungal source. Although this enzyme is potentially useful there are insufficient quantities available to allow the development of a commercial process. Work at the Massachusetts Institute of Technology has established that various other peroxidase-type enzymes may be persuaded to hydrolyse lignin. For instance, horseradish peroxidase will partially degrade natural lignin (to the same extent as the fungal ligninase) providing dioxane/water (95:5 v/v) is used as the solvent. No reaction occurs in aqueous solution. The 95% (v/v) dioxane mixture partially solubilizes the lignin and also adsorbs the free radicals produced during the reaction, thus preventing repolymerization of the polymer. The horseradish peroxidase retains only 5% of its original activity in 95% (v/v) dioxane but the enzyme is so active and so plentiful that this is not a significant problem. Other peroxidases such as lactoperoxidase are also able to work in the same way. It seems at last that there is now a way to capitalize on the degradation of lignin as a source of aromatic chemicals.

The potential of using enzymes in organic solvents is clearly an area that is ripe for development. The ability to modify hydrophobic molecules of use to the chemical industry and the introduction of novel reactions will lead to a whole range of new enzyme applications.

Synthetic enzymes

Lessons that have been learnt about the catalytic mechanisms of enzymes can be put to use in the design of novel catalysts. An enzyme, although it may contain

Future prospects

many amino acids, uses only a small proportion of these for the actual catalytic process. For instance, chymotrypsin is comprised of 245 amino-acid residues, yet it is the sidechain substituents of only three of these that have been implicated in the mechanism of proteolysis. This is perhaps an over-simplification, as other amino acids are important for forming the substrate-binding site, but it does serve to illustrate how little of the enzyme is directly involved in catalysis. One method for the design of novel catalysts is to base the structure on proteins or peptides whereas the other utilizes different classes of organic compounds. Both types of approach will be considered here.

The concept of synthesis of protein-containing enzyme analogues may be split into two sub-divisions. Either an existing enzyme or protein may be modified or a totally novel peptide structure may be constructed. The modification of enzymes has already been dealt with in detail in Chapter 9, and it is the modification of non-enzymic proteins and the construction of synthetic peptides to produce novel catalysts that will be briefly considered here.

One type of conversion of a non-enzymic protein into a selective catalyst is exemplified by the following example. Sperm-whale myoglobin, which is ordinarily an oxygen-transporting protein, has been modified to produce an oxidase-type of enzyme. A ruthenium transfer catalyst, $[Ru(NH_3)_5]^{3+}$ has been attached to each of three surface histidine residues on myoglobin and this 'bio-inorganic, semi-synthetic' enzyme is able to reduce oxygen while oxidizing various compounds such as ascorbate. The complex is nearly as effective as ascorbate oxidase for the oxidation of ascorbic acid. Other organo-metallic complexes, e.g. bis(phosphine) rhodium, are also potential candidates for the construction of semi-synthetic enzymes.

A more ambitious objective than that described above is to be able to design and build an enzyme *de novo*. Work is well underway in this area, although as stated earlier in this chapter, it is still not possible to predict accurately a three-dimensional structure from the primary sequence of a protein and this presents a major barrier to progress in this field. Currently, several groups have synthesized short peptide chains containing regions of β-pleated sheets or α-helices. Initially, the major objective has been to develop binding sites for simple organic molecules before building up to the more complex structures with catalytic activity. Some success has already been achieved but it is difficult to predict how long it will take to design from first principles and to build a totally synthetic enzyme (van Brunt, 1986).

The development of modern methods of synthesis in organic chemistry has opened up the possibility of building non-protein enzyme analogues. One of the most extensively studied systems has been based on the Schardinger dextrins (Suckling, 1984). These compounds are naturally occurring, cyclic oligosaccharides based on $\alpha 1 \rightarrow 4$-linked D-glucose residues which form crystalline solids. The three molecules of interest to the enzymologist are cyclomaltohexaose (α-cyclodextrin, Fig. 10.5), cyclomaltoheptaose (β-cyclodextrin) and cyclomaltooctaose (γ-cyclodextrin) which contain six, seven and eight glucose residues respectively. The orientation of the sugar residues is such that a cylindrical structure is formed with a relatively hydrophobic interior and with all of the

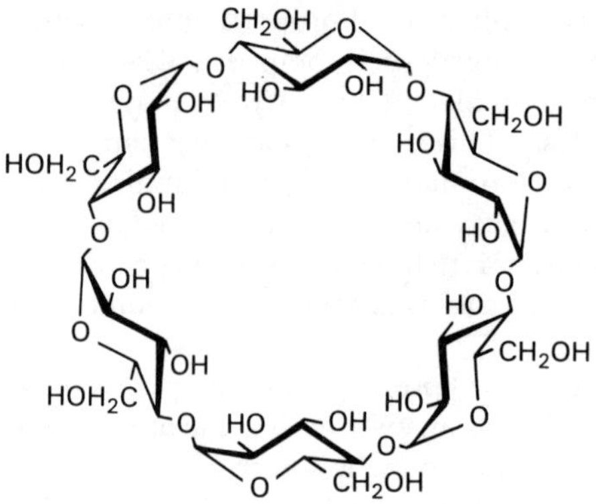

Fig. 10.5 The structure of cyclomaltohexaose (α-cyclodextrin).

hydroxyl groups extending from the two rims of the molecule. The cavities that are formed within these molecules are slightly tapered and have internal diameters of 4.5–8.5 angstroms (α to γ cyclodextrin) and a depth of 7 angstroms. This means that all three cyclodextrins can accommodate a benzene ring within the interior space but any substitution of the aromatic nucleus may prevent inclusion into the smaller of the structures. Therefore, the cyclodextrins provide a hydrophobic binding site of certain size and, with the covalent attachment of suitable reactive groups, it should be possible to produce an enzyme analogue. In fact, the sugar alcohol groups alone are able to catalyse the hydrolysis of certain esters, e.g. the *p*-nitrophenyl ester of ferrocene acrylic acid, with an enhancement of 10^5 in the rate of reaction.

The most exciting results have been obtained using β-cyclodextrin in which *O*-[4(5)-mercaptomethyl-4(5)-methylimidazol-2-yl]benzoate has been covalently attached to the rim of the molecule (Fig. 10.6). This derivative contains a

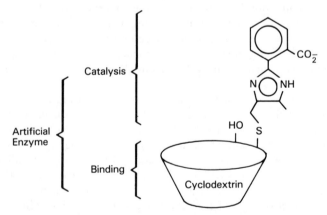

Fig. 10.6 The structure of the chymotrypsin analogue, benzyme.
(From Bender *et al.*, 1986)

Future prospects

hydroxyl group, an imidazole ring and a carboxylate group, and is a mimic of the active site of chymotrypsin. The hydroxyl group of serine-195, the imidazole ring of histidine-57 and the carboxylate ion of aspartate-102 are the residues involved in the catalytic action of chymotrypsin. The synthetic enzyme exhibits maximum activity at pH 10, and above, and is able to catalyse the hydrolysis of certain esters at least as fast as chymotrypsin. Furthermore, this artificial enzyme known as β-benzyme exhibits Michaelis–Menten kinetics and is more heat- and alkali-stable than its natural counterpart (Bender et al., 1986). Models such as β-benzyme show that it should be possible to design low molecular weight mimics for many of the simple hydrolytic enzymes provided that one has a detailed knowledge of the active site. Because of the simplicity of their structure, one might expect these cyclodextrin-based analogues to be more readily obtained and to be more stable than the natural enzymes.

The principle of using cyclic structures to provide selective binding sites has been exploited with other families of molecules. For example, the crown ethers (Fig. 10.7) and related compounds are able to bind certain chemical groups. With appropriate modification to the basic structures it is possible to develop novel binding sites together with catalytic activity. However, as yet there is not a crown-ether derivative that is analogous to the cyclodextrin-based β-benzyme.

The idea of using relatively simple organic molecules to mimic the active sites of enzymes has been shown to work and it should not be too long before we see the advent of many more semi-synthetic, enzyme-like catalysts.

Coenzyme regeneration

The majority of enzyme-catalysed reactions that are used industrially are hydrolytic. Typically, these reactions utilize one substrate (other than water) and they do not require coenzymes. For more complex reactions involving group transfers, then there will be a need to provide adequate amounts of the appropriate coenzyme (Carrea and Riva, 1986). The most common coenzymes that are required are either those involved in oxidation/reduction reactions, $NAD(P)^+/NAD(P)H$, or those participating in phosphorylation, ATP/ADP/AMP. These coenzymes are expensive to use and, unless some way is found to conserve them, industrial processes will be uneconomical. In general, attempts to separate

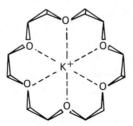

Fig. 10.7 The structure of 18-crown-6 ether.

coenzyme from the reaction products during downstream processing are less than satisfactory. Therefore, the only option is to ensure that the coenzyme is retained in an active form within the reactor. The retention of coenzyme within a reactor system entails dealing with two distinct problems. On the one hand, it is necessary physically to confine these molecules within the reaction vessel and, on the other, a process is needed to regenerate the active form of the coenzyme.

Techniques for retaining coenzyme are limited; in essence the only option is to immobilize molecules onto a high molecular weight support material. There are many methods for the immobilization of coenzymes and, for example, NAD^+ has been derivatized at the N1, the 6-amino and the C8 positions of the adenine ring and also via the ribose moiety. It transpires that derivatization through the 6-amino group of the adenine ring is the best method with both the $NAD(P)^+$ family of molecules and the ATP/ADP/AMP nucleotides. Not only is the reactivity of the coenzyme preserved but the availability of the primary amino group allows the use of conventional immobilization chemistry. These coenzymes have been immobilized on a wide range of polymeric materials but in general those that are water-soluble provide the best supports because they have only minimal effects on the enzyme-catalysed reactions. The chemistry of a typical immobilization procedure using the 6-amino position of the adenine ring is illustrated in Fig. 10.8.

The need to immobilize coenzymes on high molecular weight supports does have implications for the rest of the process. Clearly, if the enzyme is also immobilized on a support matrix, then the chances of a reaction occurring will be remote. Therefore, the use of immobilized coenzymes really dictates the immobilization method for the enzyme. Most enzyme reactors employing immobilized coenzymes have been based on ultrafiltration systems (Schmidt *et al.*, 1986). In this way both the free enzyme in solution and the high molecular weight derivative of the coenzyme may be retained within an ultrafiltration cell. Although this works very well, it does limit the number of choices that can be made regarding reactor design. An alternative method that has been considered is to immobilize coenzyme directly onto the enzyme itself (Chapter 9), although this is not really a practical alternative at the moment.

The regeneration of coenzyme can be divided into two categories: the redox, e.g. $NAD(P)^+/NAD(P)H$, and the group transfer, e.g. ATP, types of coenzyme. Examples of both types of coenzyme regeneration have been reported. With the $NAD(P)^+/NAD(P)H$ system, there are three options. Coenzyme may be regenerated chemically, electrochemically or enzymically. Chemical regeneration involves the continuous addition to the feed-stock of a compound with a suitable redox potential. Although this method can be effective it has the major drawback that the added chemical and its product will have to be removed during downstream processing. Electrochemical regeneration has also been applied successfully to model systems. However, regeneration is often limited because of the side reactions that can occur, and there are still many problems to be solved. Nonetheless, in the long term this technique seems to be very promising. Enzymic regeneration of coenzyme is, at the moment, the only practical solution. Although this means that a second enzyme must be incorporated into the reactor together

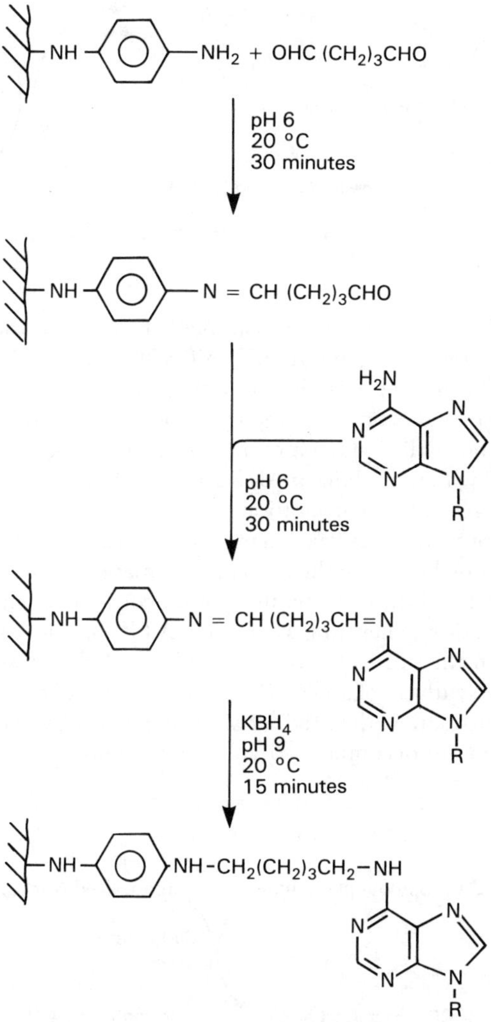

Future prospects

Fig. 10.8 An immobilization procedure for adenine-containing compounds.

with the appropriate substrate, the problems are not insurmountable. A useful enzyme for the regeneration of NADH is formate dehydrogenase. This enzyme utilizes immobilized NAD^+ and formate to produce NADH, CO_2 and hydrogen ions (Fig. 10.9). Both the CO_2 and the hydrogen ions may be readily removed from the product stream, thus preventing contamination. The formate dehydrogenase system is already in commercial use (see Chapter 7) and the prospects for other methods such as electrochemical regeneration are good in the long term.

The regeneration of ATP may be achieved only enzymically at present. Electrochemical regeneration is not an option because of the nature of the

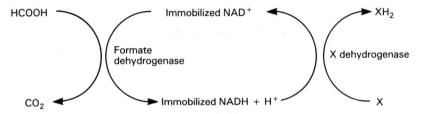

Fig. 10.9 The regeneration of NADH using formate dehydrogenase.

coenzyme and a chemical method seems unlikely. Essentially, there have been two candidate enzymes for the regeneration of ATP; these are carbamate kinase and acetate kinase. The former has the advantage that the substrate carbamoyl phosphate may be made *in situ* from cyanate and phosphate, although the equilibrium is not normally in favour of the synthesis. However, carbamate kinase does not work well with immobilized ADP and therefore the method has not been totally successful so far. Incorporation of enzyme into a reactor with free ATP/ADP has improved the utilization of this coenzyme but of course it is still lost from the reactor. The beauty of the carbamate kinase system is that the reaction product (other than ATP), carbamate, spontaneously decomposes to CO_2 and water (Fig. 10.10). A regenerating system for ATP that does work is the use of acetate kinase with the substrate acetyl phosphate. This reaction has been used successfully for the synthesis of glucose-6-phosphate using hexokinase and glucose (Fig. 10.11). Unfortunately, the product of the regeneration reaction, i.e. acetate, does not spontaneously decompose to volatile compounds as was the case with carbamate.

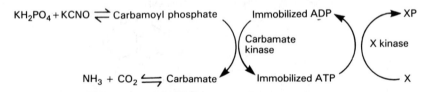

Fig. 10.10 The regeneration of ATP using carbamate kinase.

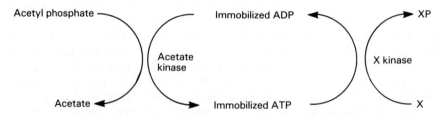

Fig. 10.11 The regeneration of ATP using acetate kinase.

Future prospects

Satisfactory methods are already available for the regeneration of both $NAD(P)^+/NAD(P)H$ and ATP coenzymes. However, these methods have drawbacks and it will not be until these are resolved that the utilization of coenzymes will be commonplace in industrial-scale, enzyme-catalysed processes.

Concluding remarks

While reading this chapter the reader will have become aware of some of the growth potentials of enzyme technology. It does not require a great stretch of the imagination to be able to focus on other areas that are probably equally as promising as those illustrated here. Whether these techniques in particular, or enzyme technology in general, will develop in the future will depend largely on the economic climate that prevails. However, what is apparent is that there is no limit to the imagination and ingenuity of scientists and engineers in their quest to develop new techniques and processes. Enzyme technology has become firmly established over the last two decades or so, and the prospects are excellent for the continued growth of the subject, both at the fundamental level and in the development of novel applications.

Appendix 1

Enzyme Commission nomenclature

The International Union of Biochemistry has established a committee (The Enzyme Commission) that is responsible for the nomenclature and classification of enzymes. Periodically, the Commission produces publications to update the list of recommended and systematic names for enzymes together with assigned Enzyme Commission numbers (EC numbers). In our book we have (within reason) used the EC recommended names for enzymes. However, we have not been hidebound by convention and occasionally we have broken the rules and used a widely accepted name rather than the recommended form, e.g. amyloglucosidase rather than the recommended glucan 1,4-α-glucosidase. Table A.1.1 contains a list of recommended and systematic names together with EC numbers for all of the enzymes mentioned in the book. A complete list of alternative names and details of the reactions that are catalysed may be derived from the publication *Enzyme Nomenclature* (Webb, 1984).

Table A1.1 Enzyme Commission Nomenclature

Recommended name[1]	Enzyme Commission number[1]	Systematic name[1]
Acetate kinase	2.7.2.1	ATP:acetate phosphotransferase
Acetylcholinesterase	3.1.1.7	Acetylcholine acetylhydrolase
Alcohol dehydrogenase	1.1.1.1	Alcohol:NAD$^+$ oxidoreductase
Alcohol oxidase	1.1.3.13	Alcohol:oxygen oxidoreductase
Alginate lyase	4.2.2.3	Poly (1,4-β-D-mannuronide) lyase
Alkaline phosphatase	3.1.3.1	Orthophosphoric-monoester phosphohydrolase (alkaline optimum)
Alkaline proteinases	3.4.21.14	Microbial serine proteinases[2]
Amidase	3.5.1.4	Acylamide amidohydrolase
Amine oxidase (copper-containing)	1.4.3.6	Amine:oxygen oxidoreductase (deaminating) (copper-containing)
L-Amino acid oxidase	1.4.3.2	L-Amino acid:oxygen oxidoreductase (deaminating)
Aminoacylase	3.5.1.14	N-Acyl L-amino acid amidohydrolase
α-Amylase	3.2.1.1	1-4-α-D-Glucan glucanohydrolase
β-Amylase	3.2.1.2	1-4-α-D-Glucan maltohydrolase
γ-Amylase *see* Glucan 1,4-α-glucosidase		
Amyloglucosidase *see* Glucan 1,4-α-glucosidase		
L-Ascorbate oxidase	1.10.3.3	L-Ascorbate:oxygen oxidoreductase
Asparaginase	3.5.1.1	L-Asparagine amidohydrolase
Aspartase *see* Aspartate ammonia-lyase		
Aspartate aminotransferase	2.6.1.1	L-Aspartate:2-oxoglutarate aminotransferase
Aspartate ammonia-lyase	4.3.1.1	L-Aspartate ammonia-lyase
Bromelain	3.4.22.4	Cysteine proteinase[2]
Carbamate kinase	2.7.2.2	ATP:carbamate phosphotransferase

Table A1.1 *(cont)*

Enzyme	EC Number	Systematic name
Catalase	1.11.1.6	Hydrogen peroxide:hydrogen peroxide oxidoreductase
Catechol oxidase	1.10.3.1	1,2-Benzenediol:oxygen oxidoreductase
Cellobiase		
see β Glucosidase		
Cellulase	3.2.1.4	1,4-(1,3;1,4)-β-D-Glucan 4-glycanohydrolase
Cholesterol esterase	3.1.1.13	Sterol-ester acylhydrolase
Cholesterol oxidase	1.1.3.6	Cholesterol:oxygen oxidoreductase
Chymosin	3.4.23.4	Aspartic proteinase[2]
Chymotrypsin	3.4.21.1	Serine proteinase[2]
Creatinase		
see Creatinine deiminase		
Creatinine deiminase	3.5.4.21	Creatinine iminohydrolase
Cytochrome P450		
α-Dextrin endo-1,6-α-glucosidase	3.2.1.41	α-Dextrin 6-glucanohydrolase
DNA-directed DNA polymerase	2.7.7.7	Deoxyribonucleoside-triphosphate:DNA deoxynucleotidyl transferase (DNA directed)
DNA-directed RNA polymerase	2.7.7.6	Nucleoside-triphosphate:RNA nucleotidyltransferase (DNA-directed)
DNA ligase		
see Polydeoxyribonucleotide synthase (ATP)		
DNA polymerase		
see DNA-directed DNA polymerase		
DNA uracil N-glycosidase		
Diamine oxidase		
see Amine oxidase (copper containing)		
Diastatic enzyme		
see α-Amylase		
Dihydropyrimidinase	3.5.2.2	5,6-Dihydropyrimidine aminohydrolase
Formate dehydrogenase	1.2.1.2	Formate: NAD^+ oxidoreductase

β-Fructofuranosidase	3.2.1.26	β-D-Fructofuranoside fructohydrolase
L-Fucose isomerase		
α-Galactosidase	3.2.1.22	α-D-Galactoside galactohydrolase
β-Galactosidase	3.2.1.23	β-D-Galactoside galactohydrolase
Galactoside acetyltransferase	2.3.1.18	Acetyl-CoA:β-D-galactoside 6-acetyltransferase
Glucan 1,4-α-glucosidase	3.2.1.3	1,4-α-D-Glucan glucohydrolase
Glucoamylase *see* Glucan 1,4-α-glucosidase		
Glucose-6-phosphate dehydrogenase	1.1.1.49	D-Glucose-6-phosphate:NADP$^+$ 1-oxidoreductase
Glucose oxidase	1.1.3.4	β-D-Glucose:oxygen 1-oxidoreductase
Glucose isomerase *see* Xylose isomerase		
α-Glucosidase *see* Glucan 1,4-α-glucosidase		
β-Glucosidase	3.2.1.21	β-D-Glucoside glucohydrolase
β-Glucuronidase	3.2.1.31	β-D-Glucuronoside glucuronoso hydrolase
Glutamate dehydrogenase	1.4.1.2	L-Glutamate:NAD$^+$ oxidoreductase (deaminating)
Hexokinase	2.7.1.1	ATP:D-hexose 6-phosphotransferase
Histidine ammonia-lyase	4.3.1.3	L-Histidine ammonia-lyase
Hyaluronidase		
Hyaluronoglucosaminidase	3.2.1.35	Hyaluronate 4-glycanohydrolase
Hydantoinase *see* Dihydropyrimidinase		
20-β-Hydroxysteroid dehydrogenase *see* (R)-20-Hydroxysteroid dehydrogenase		
(R)-20-Hydroxysteroid dehydrogenase	1.1.1.53	(20R)-17α,20,21-Trihydroxysteroid: NAD$^+$ oxidoreductase

Table A1.1 (cont)

Inulinase	3.2.1.7	2,1-β-D-Fructan fructanohydrolase
Invertase		
see β-Fructofuranosidase		
Klenow enzyme		
see DNA-directed DNA polymerase[3]		
β-Lactamase	3.5.2.6	β-Lactamhydrolase
Lactase	3.2.1.108	Lactose galactohydrolase
L-Lactate dehydrogenase	1.1.1.27	(S)-Lactate:NAD$^+$ oxidoreductase
Lactoperoxidase		
see Peroxidase		
Ligninase		
Mannan endo-1,4-βmannosidase	3.2.1.78	1,4-β-D-Mannan mannohydrolase
Methanol dehydrogenase		
see Alcohol oxidase		
Microbial aspartic proteinase	3.4.23.6	Aspartic proteinase[2]
Microbial metalloproteinase	3.4.24.4	Metallo proteinase[2]
Microbial serine proteinase	3.4.21.14	Serine proteinase[2]
NADH-specific FMN oxidoreductase		
see NAD(P)H dehydrogenase (FMN)		
NADPH-dehydrogenase	1.6.99.1	NADPH: (acceptor) oxidoreductase
NAD(P)H dehydrogenase (FMN)	1.6.8.1	NAD(P)H:FMN oxidoreductase
NADPH-specific FMN oxidoreductase		
see NAD(P)H dehydrogenase (FMN)		
Neutral proteinase (bacterial)		
see Microbial metalloproteinases		
Nitrilase	3.5.5.1	Nitrile aminohydrolase
Old yellow enzyme		
see NADPH dehydrogenase		
Pancreatic lipase		
see Triacylglyerol lipase		
Pancreatic proteinase		
see Chymotrypsin, Trypsin		

Papain	3.4.22.2	Cysteine proteinase[2]
Pectinase		
see Polygalacturonase		
Penicillin G acylase		
see Penicillin amidase		
Penicillin V acylase		
see Penicillin amidase		
Penicillin amidase	3.5.1.11	Penicillin amidohydrolase
Penicillinase		
see β-Lactamase		
Pentosanase		
see Mannan endo-1,4-β-mannosidase		
Pepsin	3.4.23.1	Aspartic proteinase[2]
Peroxidase	1.11.1.7	Donor:hydrogen-peroxide oxidoreductase
Plasminogen activator	3.4.21.31	Serine proteinase[2]
Polydeoxyribonucleotide synthase (ATP)	6.5.1.1	Poly (deoxyribonucleotide):poly (deoxyribonucleotide) ligase (AMP-forming)
Polygalacturonase	3.2.1.15	Poly (1,4-α-D-galacturonide) glycanohydrolase
Polyphenol oxidase		
see Catechol oxidase		
Pullulanase		
see α-Dextrin endo-1,6 α-glucosidase		
Rennin (calf)		
see Chymosin		
Rennin (Mucor)		
see Microbial aspartic proteinases		
Restriction endonuclease		
see Type II site-specific deoxyribonuclease		
Reverse transcriptase		
see RNA-directed DNA polymerase		

Table A1.1 *(cont)*

Ribonuclease H	3.1.26.4	Endoribonucleases producing 5'-phosphomonoesters[2]
RNA-directed DNA polymerase	2.7.7.49	Deoxynucleoside-triphosphate: DNA deoxynucleotidyl transferase (RNA-directed)
Subtilisin *see* Microbial serine proteinases		
Thiogalactoside transacetylase *see* Galactoside acetyltransferase		
Triacylglycerol lipase	3.1.1.3	Triacylglycerol acylhydrolase
Trypsin	3.4.21.4	Serine proteinase[2]
Type II restriction endonuclease *see* Type II site-specific deoxyribonuclease		
Type II site-specific deoxyribonuclease	3.1.21.4	Endodeoxyribonucleases producing 5'-phosphomonoesters[2]
Urease	3.5.1.5	Urea amidohydrolase
Urate oxidase	1.7.3.3	Urate:oxygen oxidoreductase
Uric acid oxidase *see* Urate oxidase		
Uricase *see* Urate oxidase		
Urokinase *see* Plasminogen activator		
Xylose isomerase	5.3.1.5	D-Xylose ketol-isomerase

1. This information has been derived from the book entitled Enzyme Nomenclature, (Webb 1984).
2. These enzymes do not have individual systematic names. Instead the class of reaction or type of catalysis has been indicated.
3. The Klenow enzyme is a fragment of DNA-directed DNA polymerase I obtained after limited proteolysis.

Appendix 2

Residence time distribution analysis

When a stream of material flows through a reactor vessel, it is usual to base performance calculations on one of two assumptions:

(1) that the fluid in the reactor is ideally mixed, such that its composition is uniform and is equivalent to the composition of the product stream (continuous stirred tank reactor, CSTR)
(2) elements of fluid that enter the reactor at the same moment move through at a constant velocity and leave at the same moment (plug flow reactor, PFR).

These reactor types, which are discussed in Chapter 4, represent ideal cases but, in practice, there will be a compromise, i.e. some back-mixing in a PFR and some dead zones in a CSTR. It may be important to quantify this discrepancy from ideal behaviour, especially when a new reactor type is considered.

The F diagram

If we define the volume of the reactor vessel occupied by fluid to be V and the volumetric flow rate through the reactor to be Q, we can make some component of the inlet flow undergo a sudden step change in concentration (e.g. introduce a dye). By following the change of concentration of this component in the exit stream, we can assess the degree of mixing within the reactor. The fraction of the dye material in the outflow (F) at a give time (t) after injection can be described as $F(t)$. The plot of $F(t)$ against $(Qt)/V$ is known as an F diagram. The characteristic curves can be seen in Fig. A.2.1. The shape of the F diagram depends on the relative times taken by various fluid elements to pass through the reactor vessel, i.e. the distribution of residence times.

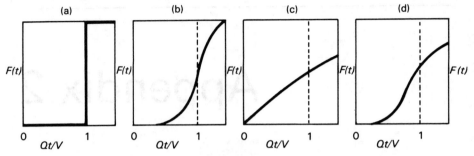

Fig. A.2.1 Characteristic F diagrams for (a) plug flow, (b) plug flow with some axial mixing, (c) perfect mixing, (d) mixing with some dead zones.

Quantifying the deviation from an ideal state

Taking the ideal situation of a single well-stirred tank, for a given step change in concentration of a tracer in the feed stream (C_o), the exit concentration (C_i) is described by

$$QC_i + V\frac{dc}{dt} = QC_o$$

This can be integrated to give

$$\frac{C_i}{C_o} = 1 - \exp(-Qt/V)$$

This expression gives the fractional response with time. So a plot in $\ln F$ against t will have a slope of Q/V and an intercept of zero.

If we now consider the deviations which can arise from ideal mixing, there are two simple cases which can occur.

(a) *Single well-stirred vessel with short circuiting and partial stagnation* This is shown schematically in Fig. A.2.2 and considers the situation where a fraction of the feed (f_1) enters the mixing zone but the remainder passes straight through. Of the material entering the mixing zone (f_2), a fraction becomes trapped in a stagnant zone. The equation describing the mixing zone is

$$f_1 Q C_o = f_1 Q C + f_2 V \frac{dc}{dt}$$

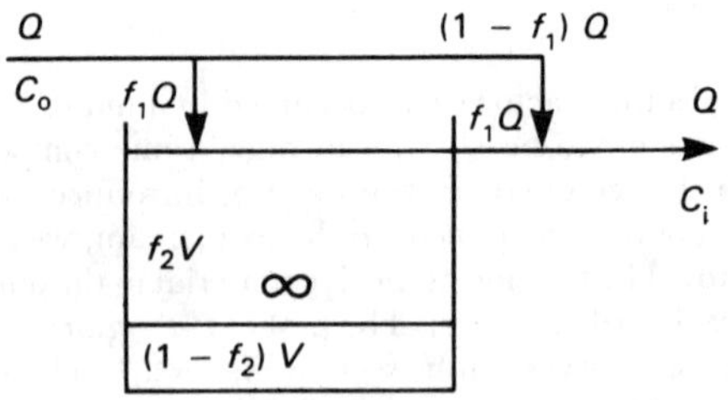

Fig. A.2.2 Diagram of reactor where both bypassing and dead zones occur.

Residence time distribution analysis

which can be integrated to give

$$\frac{C_i}{C_o} = \exp\left(-f_1 Qt/f_2 V\right)$$

This can be combined with the outflow stream equation to give

$$F(t) = (1 - f_1) + \exp\left(-f_1 Qt/f_2 V\right)$$

We can now plot $\ln[F(t)]$ against Qt/V to calculate the values of the parameters. The intercept is $\ln[F(t)]$ and the slope is $-f_1/f_2$. Obviously, in the case where $f_1 = 1 = f_2$, the reactor is well-mixed.

(b) *Reactor is a compromise between well-mixed and plug flow* In this case a fraction of the hold-up volume is considered to be well-mixed. The throughput Q flows in series through the remaining volume with the same residence-time distribution being seen whichever region is encountered first (Fig. A.2.3). The composite equation for this situation can be derived and shown to be

$$\frac{C_i}{C_o} = \exp\left(\frac{-1}{f_2} \cdot \frac{Qt}{V} - (1 - f_2)\right)$$

In this case, a plot of $\ln[F]$ against $(Qt)/V$ allows the calculation of f_2 from both slope and intercept, where f_2 is the fraction of the reactor operating in the well-mixed mode.

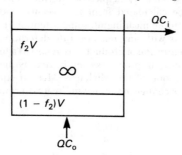

Fig. A.2.3 Diagram of reactor showing plug flow and well-mixed regions in series.

Summary

The approach described here represents just one of a number of tests to evaluate reactor mixing. This assessment may be particularly appropriate where an intermediate reactor type (possibly a fluidized bed) or a recycle reactor is being considered.

Appendix 3

Design of enzyme assays

In Chapter 4 we discussed some principles of enzyme kinetics and showed how the Michaelis constants can be calculated from rate versus reactant concentration data. However, if these data are to be meaningful, some attention must be paid to the way in which they are obtained. As with any reaction rate determination, we must be able to monitor the change in concentration of product or reactant as a function of time. In the case of enzyme catalysed reactions, this process is complicated by the nature of these catalysts.

In Chapter 4 our derivation of the Michaelis–Menten equation was based on two assumptions. First, the rate of change in the enzyme/reactant concentration was assumed to be zero, i.e.

$$\frac{d[ER]}{dt} = 0$$

In practice, we usually need to measure the initial rate. At this point the concentration of ER will represent the true value for the concentration of the reactant R, provided

$$[ER] = \frac{[E][R]}{K_m}$$

where $[E]$ is the concentration of free enzyme. Obviously, as the reaction proceeds, the concentration of R will decrease, leading to a lower rate.

The second assumption made is that we can construct a mass balance of the enzyme in the reaction such that

$$[E_{tot}] = [E] + [ER]$$

If the value of $[E_{tot}]$ changes throughout the reaction, then we will obtain anomalous results. While the enzyme protein concentration cannot change, it may be that the fraction of enzyme which is active will fall as a result of denaturation. To obtain realistic data for a determination of V_{max} and K_m we must obtain accurate estimates of the reaction rate at zero

Design of enzyme assays

time. As we cannot determine rate from a single point, we must extrapolate for some later time point(s). If we are to do this accurately, we must know that the active enzyme concentration has not changed during this period. This can be relatively easily confirmed using a test proposed by Selwyn in 1965.

If we consider product concentration as a function of $[E_{tot}]$ multiplied by reaction time, the experimental conditions could be arranged such that product concentration should be identical for the effects of high enzyme concentration over a short time and a low enzyme concentration over a long time. This would be true, providing that the enzyme is stable over the time course of the reaction. If the enzyme is unstable, then the reaction utilizing the smaller amount of enzyme will show a lower conversion. This can be represented graphically (Fig. A.3.1). Having established the stability of the enzyme during the course of the assay, the next step is to obtain a progress curve to allow the determination of the initial rate.

In considering an enzyme assay, a choice on the approach to be used may be necessary. In some cases it will be possible to monitor the reaction continuously by using a spectrophotometer or other sensor, which allows the progress curve to be plotted directly onto a chart recorder. If reactants or products cannot be measured directly, it may be possible to couple the reaction to a second conversion giving a product which can be assayed. Providing that the primary reaction represented the rate-limiting step, the appearance of the second product would represent the true rate of reaction. Where direct assays of this sort are possible, it is relatively easy to obtain a suitable progress curve simply by varying the enzyme concentration used. The reactant concentrations will be dictated by the kinetics of the enzyme and may have to be established by an initial trial experiment (Chapter 4).

When it is not possible to construct a direct assay, we have to resort to an indirect method where a sample is removed from the reaction mixture. The reaction is stopped and the product/reactant concentration is determined using conventional analytical techniques. Compared with a direct assay this approach is more tedious but, more significantly, the time constraints make obtaining data close to zero time extremely difficult. So we now have the problem of determining initial rate from a set of data where the first point may be taken a minute or more after the reaction was initiated (Fig. A.3.2). In this case simply

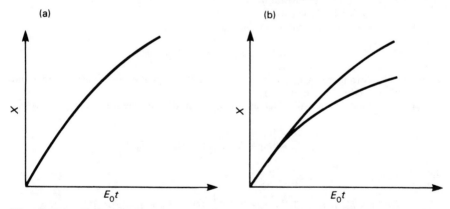

Fig. A.3.1 Selwyn's test for enzyme inactivation: X (fractional conversion) against $E_0 t$ (arbitary units). (a) Data obtained at different E_0 values fall on a common line, showing no inactivation. (b) Data obtained at different E_0 values fall on separate lines, showing the effects of inactivation where smaller amounts of enzyme are incubated for longer periods.

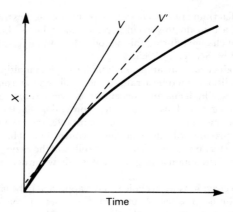

Fig. A.3.2 Plot of fractional conversion as a function of time, showing the true value of v and the error introduced by taking the wrong asymptote (v').

taking a tangent to the curve at time zero may lead to significant errors in our estimate of v. An important factor here is that enough points should be determined to enable a realistic initial rate to be determined.

To avoid the errors inherent in drawing a line of eye, methods have been devised based on fitting the data obtained to an integrated rate equation. One such approach is based on the principle of the direct linear plot (Chapter 4). Consider the integrated Michaelis-Menten equation

$$V_{max} \cdot t = [R_0] - [R] + K_m \ln\{[R_0]/[R]\}$$

where V_{max} = maximum velocity
$[R_0]$ = reactant concentration at time zero
$[R]$ = reactant concentration
K_m = the Michaelis constant

This can be written in terms of product as

$$V_{max} \cdot t = [P] + K_m \ln\{[P_{eq}]/([P_{eq}] - [P])\}$$

where $[P_{eq}]$ = product concentration at equilibrium ($=[R_0]$ for an irreversible reaction)
$[P]$ = product concentration

So for any time t_i there will be a corresponding product concentration $[P_i]$. Therefore,

$$V_{max} \cdot t_i = [P_i] + K_m \ln\{[P_{eq}]/([P_{eq}] - [P_i])\}$$
$$V_{max} \cdot t_j = [P_j] + K_m \ln\{[P_{eq}]/([P_{eq}] - [P_j])\}$$

Therefore

$$\frac{1}{t_i}([P_i] + K_m \ln\{[P_{eq}]/([P_{eq}] - [P_i])\}) = \frac{1}{t_j}([P_j] + K_m \ln\{[P_{eq}]/([P_{eq}] - [P_j])\})$$

This can be solved for K_m:

$$K_{mij} = \frac{(t_j[P_i] - [P_j]t_i)}{(t_i \ln\{[P_{eq}]/([P_{eq}] - [P_j])\}) - t_j \ln\{[P_{eq}]/([P_{eq}] - [P_i])\}}$$

Design of enzyme assays

The corresponding value of V_{maxij} is obtained by substituting the value of K_m back into the original integrated rate equation.

$$V_{\text{maxij}} = \frac{([P_i] + K_{\text{mij}} \ln\{[P_{\text{eq}}]/([P_{\text{eq}}] - [P_i])\})}{t_i}$$

These values can be obtained for each data pair obtained.

Now, we have already discussed the use of the integrated Michaelis–Menten equation for the determination of K_m and V_{max} (Chapter 4) and concluded that it is not an ideal method. However, if we take the estimated values and substitute them back into the Michaelis–Menten equation, we can obtain an estimate for the initial rate, given the substrate concentration, i.e.

$$V_{ij} = (V_{\text{maxij}} [P_{\text{eq}}])/(K_{\text{mij}} + [P_{\text{eq}}])$$

By taking the median of the v values, we obtain the best estimate of the initial rate. The author of this method has shown that although the values of K_m and V_{max} obtained are sensitive to errors in the assumptions, the estimates of initial rates are remarkably insensitive to similar errors.

Although both this and the direct linear plot method can be carried out graphically, it is generally more convenient to use a microcomputer to solve the resulting equations.
The listings of suitable programs written in BASIC are provided.

```
1000 REM ***  DLP...J.HUBBLE 1986 ***
1010 REM
1020 REM   Main variables
1030 REM
1040 REM   S(1-N)    Array of substrate values
1050 REM   V(1-N)    Array of rate values
1060 REM   B1(1-L)   Array of 1/Vmax values
1070 REM   A1(1-L)   Array of Km/Vmax values
1080 REM   G,F,M     Loop counters
1090 REM   U,Q       Exchange variables for sort
1100 REM   Y,Z       Hold final values for output
1110 REM
1120 CLS
1130 PRINT"THIS PROGRAM CALCULATES KM & VMAX"
1140 PRINT"USING THE MODIFIED DIRECT LINEAR PLOT"
1150 PRINT"Cornish-Bowden & Eisenthal (1978) B.B.A. 523, 268"
1160 PRINT"MAXIMUM NO OF DATA PAIRS =(10)"
1170 PRINT
1180 DIM S(10),V(10),A1(45),B1(45)
1190 PRINT"ENTER NUMBER OF DATA PAIRS"
1200 REM
1210 REM Data input
1220 REM
1230 INPUT N
1240 PRINT"ENTER S&V VALUES SEPERATED BY A COMMA"
1250 FOR I=1 TO  N
1260 INPUT S(I),V(I)
1270 NEXT I
1280 CLS
1290 REM
1300 REM Set counter for loops
1310 REM
1320 M=N-1
1330 F=2
1340 G=F
1350 L=1
1360 REM
1370 REM Calculate values of 1/Vmax and Km/Vmax for every
1380 REM pair combination of S & V values
1390 REM
1400 PRINT"EACH VALUE SHOULD BE COMPARED WITH EVERY OTHER"
1410 FOR I=1 TO M
1420 FOR J=G  TO N
1430 PRINT"I"I,"J"J
1440 P=(V(I)-V(J))
1450 B1(L)=(S(J)/V(J)-S(I)/V(I))/(S(J)-S(I))
1460 A1(L)=S(I)/V(I)-(S(I)/V(I))*B1(L)/(1/V(I))
1470 L=L+1
1480 NEXT J
1490 F=F+1
1500 G=F
1510 NEXT I
1520 PRINT
1530 REM
1540 REM Set counters for sorting routine
1550 REM
1560 F=2
1570 G=F
1580 L=L-1
1590 REM
```

Design of enzyme assays

```
1600 REM   Set counters for sorting routines
1610 REM
1620 F=2
1630 G=F
1640 L=L-1
1650 REM
1660 REM   Bubble sort for intermediate values of Km, Vmax, and V
1670 REM
1680 FOR I=1 TO L-1
1690 FOR J=G TO L
1700 IF A1(I) >=A1(J) GOTO 1740
1710 U=A1(I)
1720 A1(I)=A1(J)
1730 A1(J)=U
1740 IF B1(I) >=B1(J) GOTO 1780
1750 Q=B1(I)
1760 B1(I)=B1(J)
1770 B1(J)=Q
1780 IF V(I) >= V(J) GOTO 1820
1790 R=V(I)
1800 V(I)=V(J)
1810 V(J)=R
1820 NEXT J
1830 F=F+1
1840 G=F
1850 NEXT  I
1860 REM
1870 REM   Print out of sorted intermediate values (optional)
1880 REM
1890 PRINT "SORTED VM & KM & V VALUES"
1900 PRINT
1910 FOR I=1 TO L
1920 PRINT B1(I),A1(I),V(I)
1930 NEXT I
1940 REM
1950 REM   Determine median values of intermediate arrays
1960 REM
1970 M=L
1980 M=M/2
1990 L=L/2+.4
2000 L=INT(L)
2010 PRINT
2020 IF (M-L)<=.1 GOTO 2120
2030 REM
2040 REM   Number of values is odd - take median point
2050 REM
2060 Z=M+(M-L)
2070 PRINT "VMAX="B1(Z),"KM="A1(Z),"V="V(Z)
2080 GOTO 2210
2090 REM
2100 REM   Number of vales is even - take average of two median points
2110 REM
2120 Y=(B1(L)+B1(L+1))/2
2130 Z=(A1(L)+A1(L+1))/2
2140 ZZ=(V(L)+V(L+1))/2
2150 PRINT "VMAX="Y,"KM="Z ,"V="ZZ
2160 REM
2170 REM   N.B. Km and Vmax values are included for diagnostic
2180 REM   purposes only.  Values of these parameters determined
2190 REM   from a single rate curve are very unreliable.
2200 REM
2210 END
```

```
1000 REM ***   INITIAL RATE ...J.HUBBLE 1986 ***
1010 REM
1020 REM
1030 REM     Main Variables
1040 REM
1050 REM     T(1-N)   Array of time values
1060 REM     P(1-N)   Array of product values
1070 REM     A1(1-L)  Array of Km values
1080 REM     B1(1-L)  Array of Vmax values
1090 REM     V(1-L)   Array of V values
1100 REM     G,F,M    Loop counters
1110 REM     U,Q,R    Exchange variables for sorts
1120 REM     Y,Z,ZZ   Hold final vales for output
1130 REM     PEQ      Product concentrate at equilibrium point of
1140 REM              the reaction (can be taken as the product
1150 REM              concentration where the curve flattens out)
1160 CLS
1170 PRINT "THIS PROGRAM CALCULATES INITIAL RATES FROM BATCH ASSAYS"
1180 PRINT "USING A MODIFIED DIRECT LINEAR PLOT"
1190 PRINT "Cornish-Bowden (1975) B.J. 149, 305"
1200 REM
1210 REM   Data input section
1220 REM
1230 PRINT "MAXIMUM NO OF DATA PAIRS =(10)"
1240 PRINT
1250 DIM T(10),P(10),A1(45),B1(45),V(45)
1260 PRINT "ENTER NUMBER OF DATA PAIRS"
1270 INPUT N
1280 PRINT "ENTER T & P VALUES SEPERATED BY A COMMA"
1290 FOR I=1 TO N
1300 INPUT T(I),P(I)
1310 NEXT   I
1320 INPUT"P EQULILIBRIUM";PEQ
1330 REM
1340 CLS
1350 REM
1360 REM   Set counters for nested loops
1370 REM
1380 M=N-1
1390 F=2
1400 G=F
1410 L=1
1420 REM
1430 REM   Calculate intermediate Km, Vmax, and V values
1440 REM   for every pair combination of T & P
1450 REM
1460 PRINT "EACH VALUE SHOULD BE COMPARED WITH EVERY OTHER"
1470 FOR I=1 TO M
1480 FOR J=G TO N
1490 PRINT "I"I,"J"J
1500 A1(L)=(T(I)*P(J)-T(J)*P(I))/(T(J)*LOG(PEQ/(PEQ-P(I)))-T(I)*LOG(PEQ/(PEQ-P(J))))
1510 B1(L)=(P(I)+A1(L)*LOG(PEQ/(PEQ-P(I))))/T(I)
1520 V(L)=(B1(L)*PEQ)/(A1(L)+PEQ)
1530 L=L+1
1540 NEXT   J
1550 F=F+1
1560 G=F
1570 NEXT   I
1580 PRINT
1590 REM
```

Design of enzyme assays

```
1600 REM Bubble sort routines
1610 REM
1620 FOR I=1 TO  L-1
1630 FOR J=G TO  L
1640 IF A1(I) >=A1(J) GOTO  1680
1650 U=A1(I)
1660 A1(I)=A1(J)
1670 A1(J)=U
1680 IF B1(I) >=B1(J) GOTO  1720
1690 Q=B1(I)
1700 B1(I)=B1(J)
1710 B1(J)=Q
1720 NEXT J
1730 F=F+1
1740 G=F
1750 NEXT I
1760 REM
1770 REM Print out of sorted values (optional)
1780 REM
1790 PRINT"SORTED VM & KM INTERCEPTS"
1800 PRINT
1810 FOR I=1 TO  L
1820 PRINT B1(I),A1(I)
1830 NEXT I
1840 REM
1850 REM Determine median values of intermediate arrays
1860 REM
1870 M=L
1880 M=M/2
1890 L=L/2+.4
1900 L=INT(L)
1910 PRINT
1920 IF (M-L)<=.1 GOTO  2050
1930 REM
1940 REM Number of values is odd - take median point
1950 REM
1960 Z=M+(M-L)
1970 B1(Z)=1/B1(Z)
1980 A1(Z)=A1(Z)*B1(Z)
1990 PRINT"VMAX="B1(Z),"KM="A1(Z)
2000 GOTO  2140
2010 REM
2020 REM Number of values is even - take mean of the two
2030 REM median points
2040 REM
2050 Y=(B1(L)+B1(L+1))/2
2060 Z=(A1(L)+A1(L+1))/2
2070 Y=1/Y
2080 Z=Z*Y
2090 PRINT"VMAX="Y,"KM="Z
2100 REM
2110 REM N.B. the modified version of the DLP avoids the
2120 REM problems of intersections in the third quadrant
2130 REM
2140 END
```

References

Units and Symbols

Perry, R.H. and Green, D. (Eds) (1984). *Perry's Chemical Engineering Handbook*, 6th edition. New York, McGraw Hill.

Chapter 1

Atkinson, B. (1974). *Biochemical Reactors*. London, Pion Ltd.
Bjurstrom, E. (1985). Biotechnology, *Chemical Engineering* 126–158.
Dunnill, P. (1980). The current status of enzyme technology, in *Enzymic and Non-Enzymic Catalysis*. Eds Dunnill, P., Wiseman, A. and Blakebrough, N., pp. 28–53. Chichester, Ellis Horwood.
Harnisch, H. and Wohner, G. (1985). The importance of biotechnology for industrial chemistry, *German Chemical Engineering* **8,** 139–146.
Katchalski-Katzir, E. and Freeman, A. (1982). Enzyme engineering reaching maturity, *Trends in Biochemical Sciences* **7,** 427–431.
Klibanov, A. (1983). Immobilised enzymes and cells as practical catalysts, *Science* **219,** 722–727.
Lewis, C. and Kristiansen, B. (1985). Chemicals manufacture via biotechnology—the prospects for Western Europe, *Chemistry and Industry* 571–576.
Lilly, M.D. (1977). A comparison of cells and enzymes as industrial catalysts, in *Biotechnological Applications of Proteins and Enzymes*. Eds Bohak, Z. and Sharon, N., pp. 127–140. New York, Academic Press.
Reichelt, J.R. (1983). Toxicology in *Industrial Enzymology*. Eds Godfrey, A. and Reichelt, J. pp. 138–156. Byfleet, The Nature Press.

Chapter 2

Booth, I.R. and Higgins, C.F. (Eds) (1986). *Regulation of Gene Expression—25 Years on.* Cambridge, Cambridge University Press.
Davis, B.D. and Tai, P-C. (1980). The mechanism of protein secretion across membranes, *Nature* **283**, 433–438.
deDuve, C. (1985). *A Guided Tour of the Living Cell.* New York, Scientific American Books.
Freedman, R.B. and Hawkins, H.C. (Eds) (1980). *The Enzymology of Post-Translational Modification of Proteins,* Vol. 1. London, Academic Press.
Godfrey, T. and Reichelt, J.R. (1983). Introduction to industrial enzymology, in *Industrial Enzymology.* Eds Godfrey, A. and Reichelt, J., pp. 1–7. Byfleet, The Nature Press.
Holland, I.B., Mackman, N. and Nicaud, J-M. (1986). Secretion of proteins from bacteria, *Biotechnology* **4**, 427–431.
Old, R.W. and Primrose, S.B. (1985). *Principles of Gene Manipulation,* 3rd edition. Oxford, Blackwell Scientific.
Walter, P., Gilmore, R. and Blobel, G. (1984). Protein translocation across the endoplasmic reticulum, *Cell* **38**, 5–8.
Webb, E.C. (Ed.) (1984). *Enzyme Nomenclature, 1984.* London, Academic Press.

Chapter 3

Bonnerjea, J., Oh, S., Hoare, M. and Dunnill, P. (1986). Protein purification: the right step at the right time, *Biotechnology* **4**, 954–958.
Chase, H.A. (1984). Affinity separations utilizing immobilized monoclonal antibodies—a new tool for the biochemical engineer, *Chemical Engineering Science* **39**, 1099–1125.
Colowick, S.P. and Kaplan, N.O. (Eds) (1971). Enzyme purification and related techniques, in *Methods in Enzymology,* Vol. 22. London, Academic Press.
Dean, P.D.G., Johnson, W.S. and Middle, F.A. (Eds) (1985). *Affinity Chromatography—A Practical Approach.* Oxford, IRL Press.
Dunnill, P. (1983). Trends in downstream processing of proteins and enzymes, *Process Biochemistry* **18**, 9–13.
Dwyer, J.L. (1984). Scaling up of bio-product separation with high performance liquid chromatography, *Biotechnology* **2**, 957–964.
Kroner, K.H., Hustedt, H. and Kula, M.R. (1984a). Extractive enzyme recovery: economic considerations, *Process Biochemistry* **19**, 170–179.
Kroner, K.H., Schutte, H., Hustedt, H. and Kula, M-R. (1984b). Cross-flow filtration in the downstream processing of enzymes, *Process Biochemistry* **19**, 67–74.
Scopes, R.K. (1982). *Protein Purification: Principles and Practice.* New York, Springer Verlag.
Waterfield, M.D. (1986). Separation of mixtures of proteins and peptides by high performance liquid chromatography, in *Practical Protein Chemistry.* Ed. Darbee, A., pp. 181–205. Chichester, Wiley & Sons.

Chapter 4

Buchholz, K., Ehrenthal, E., Gloger, M., Hennrich, N., Jaworek, D., Kasche, V., Klein, J., Kramer, D.M., Kula, M.R., Manecke, G., Palm, D., Schlunsen, J. and Wagner, F. (1979). 'Characteristic properties and summary of determination methods, in *Characterisation of Immobilized Biocatalysts.* Ed. Buchholz, K., pp. 1–48. Weinheim, Verlag Chemie.

Cornish-Bowden, A. (1979). *Fundamentals of Enzyme Kinetics*. London, Butterworth.
Eisenthal, R. and Wharton, C. (1981). *Molecular Enzymology*. London, Blackie.
Eisenthal, R. and Cornish-Bowden, A. (1974). The direct linear plot, *Biochemical Journal* **139**, 715–720.
Fersht, A. (1977). *Enzyme Structure and Mechanism*. Reading, W.H. Freeman.
Morris, J.G. (1974). *A Biologists Physical Chemistry*. London, Edward Arnold.
Palmer, T. (1981). *Understanding Enzymes*. Chichester, Ellis Horwood.
Vieth, W.R., Venkatasubramanian, K., Constinides, A. and Davidson, B. (1976). Design and analysis of immobilized-enzyme flow reactors, in *Applied Biochemistry and Bioengineering*. Eds Wingard, L.B., Katchalski-Katzir, E. and Goldstein, C., pp. 221–327. New York, Academic Press.
Vieth, W.R. and Venkatasubramanian, K. (1973). Enzyme engineering—IV: Process engineering for immobilised enzyme systems, *Chemtech* **1**, 30–40.
Weetall, H.H. and Pitcher, W.H. (1986). Scaling up an immobilised enzyme system, *Science* **232**, 1396–1403.

Chapter 5

Beutler, E. (1981). Enzyme replacement therapy, *Trends in Biochemical Sciences* **6**, 95–97.
Chang, T.M.S. (1977) Artificial kidney, artificial liver, and detoxifier based on artificial cells, immobilized proteins and immobilized enzymes, in *Biomedical Applications of Immobilized Enzymes and Proteins*. Ed. Chang, T.M.S. pp. 281–295, London, Plenum Press.
Cooney, D.A. and Rosenbluth, R.J. (1975). Enzymes as therapeutic agents, *Advances in Pharmacology and Chemotherapy* **12**, 185–289.
Flint, E.J., DeGiovanni, J., Cadigan, P.J., Lamb, P. and Pentecost, B.L. (1982). Effect of GL enzyme (a highly purified form of hyaluronidase) on mortality after myocardial infarction, *Lancet* **i**, 871–874.
Goodchild, M.C. and Dodge, J.A. (1985). *Cystic Fibrosis: Manual of Diagnosis and Management*, 2nd edition, pp. 92–97. London, Baillière Tindall.
Hers, H.G. and de Barsy, T. (1973) Type II glycogenosis (acid maltase deficiency) in *Lysosomes and Storage Diseases*. Eds Hers, H.G. and van Hoof, F., pp. 197–216. London, Academic Press.
Lagerlof, E., Nathorst-Westfelt, B., Eckstrom, B. and Sjoberg, B. (1976). Production of 6-amino penicillanic acid with immobilized *Escherichia coli* acylase, in *Methods in Enzymology*. Ed. Mosbach, K., Vol. 44, pp. 759–768. London, Academic Press.
Maciag, T., Mochan, B., Kelly, P., Pye, E.K. and Iyengar, M.R. (1977) Plasminogen activators for therapeutic applications, in *Biomedical Applications of Immobilized Enzymes and Proteins*. Ed. Chang, T.M.S., Vol. 2, pp. 303–316, London, Plenum Press.
Sofer, S.S. (1979). Hepatic microsomal enzymes: potential applications, *Enzyme and Microbial Technology* **1**, 3–8.
Tager, J.M. (1985). Biosynthesis and deficiency of lysosomal enzymes, *Trends in Biochemical Sciences* **10**, 324–326.
Voller, A., Bidwell, D.E. and Bartlett, A. (1979). *The Enzyme Linked Immunosorbant Assay (E.L.I.S.A.)*, Vols. 1 and 2. Billinghurst, Dynatech Laboratories.

Chapter 6

Atkinson, B. (1985). Immobilised cells their application and potential, in *Process Engineering Aspects of Immobilised Cell Systems*. Eds Webb, C., Black, G.M. and Atkinson, B., pp. 3–34. Rugby, I.Chem.E.

Buchholz, K. (1982). Reaction engineering parameters for immobilized biocatalysts, in *Advances in Biochemical Engineering*. Ed. Fiechter, A., Vol. 24, pp. 39–71. Berlin, Springer Verlag.
Cheetham, P.S.J. (1983). The application of immobilised cells and biochemical reactions in biotechnology—principles of enzyme engineering, in *Principles of Biotechnology*. Ed. Wiseman, A., pp. 172–208. Guildford, Surrey University Press.
Cornish-Bowden, A. (1979). *Fundamentals of Enzyme Kinetics*, London, Butterworth.
Furui, M. and Yamashita, K. (1985). Diffusion coefficients in immobilised cell catalysts, *Journal of Fermentation Technology* **63**, 167–173.
Goldstein, L. (1976). Kinetic behaviour of immobilised enzyme systems, in *Methods in Enzymology*. Ed. Mosbach, K., Vol. 44, pp. 397–443. New York, Academic Press.
Goldstein, C. and Katchalski-Katzir, E. (1976). Immobilised enzymes—a survey, in *Applied Biochemistry and Bioengineering*. Eds. Wingard, L.B., Katchalski-Katzir, E. and Goldstein, L., Vol. 1, pp. 1–22. New York, Academic Press.
Hannoun, B.J.M. and Stephanopoulos, G. (1986). Diffusion coefficients of glucose and ethanol in cell free and cell occupied calcium alginate, *Biotechnology and Bioengineering* **28**, 829–835.
Horvath, C. and Engasser, J-M. (1974). External and internal diffusion in heterogeneous enzyme systems, *Biotechnology and Bioengineering* **16**, 909–923.
McCabe, W.L., Smith, J.C. and Harriott, P. (1985). *Unit Operations of Chemical Engineering*. New York, McGraw Hill.
Messing, R.A. (1985) Immobilization techniques—enzymes, in *Comprehensive Biotechnology*. Ed. Moo-Young, M., Vol. 2, pp. 191–201. Oxford, Pergamon.
Trevan, M.D. (1980). *Immobilised Enzymes*. Chichester, John Wiley.
Vieth, W.R. and Venkatasubramarian, K. (1973) Enzyme engineering—III: Properties of immobilised enzyme systems, *ChemTech*. **1**, 18–29.
Wiseman, A. (1978) Stabilisation of enzymes, in *Topics in Enzyme and Fermentation Biotechnology*. Ed. Wiseman, A., Vol. 2, pp. 280–303. Chichester, Ellis-Horwood.
Woodward, J. (Ed.) (1985) *Immobilised Cells and Enzymes: A Practical Approach*. Oxford, IRL Press.

Chapter 7

Adler-Nissen, J. (1986). *Enzymic Hydrolysis of Food Proteins*. Barking, Elsevier Applied Science.
Antrim, R.C., Kolilla, W. and Schnyder, B.J. (1979). Glucose isomerase production of high fructose syrups, in *Applied Biochemistry and Bioengineering*. Eds Wingard, L.B., Katchalski-Katzir, E. and Goldstein, L., Vol. 2, pp. 97–155, New York, Academic Press.
Barfoed, H.C. (1983). Detergents, in *Industrial Enzymology*. Eds Godfrey, A. and Reichelt, J., pp. 284–293. Byfleet, The Nature Press.
Beppu, T. (1983). The cloning and expression of chymosin genes in microorganisms, *Trends in Biotechnology* **1**, 85–89.
Coker, L.E. and Venkatasubramanian, H.J. (1985). Starch conversion processes, in *Comprehensive Biotechnology*. Ed. Moo-Young, M., Vol. 3, 777–778. Oxford, Pergamon Press.
Daniels, M.J. (1985). An industrial scale immobilised enzyme system, in *Process Engineering Aspects of Immobilised Cell Systems*. Eds Webb, C., Black, G.M. and Atkinson, B. Rugby, I.Chem.E.
Gaden, E.L., Mandels, M.H., Reese, E.T. and Spano, L.A. (Eds) (1976). Enzymatic conversion of cellulosic materials: technology and applications, in *Biotechnology and Bioengineering Symposium* 6. New York, John Wiley.

Gekuas, V., Lopez-Leiva, M. (1985) Hydrolysis of lactose: a literature review, *Process Biochemistry* **20**, 2–12.
Godfrey, A. (1983) Brewing, in *Industrial Enzymology*. Eds. Godfrey, A. and Reichelt, J., pp. 221–259. Byfleet, The Nature Press.
Klibanov, A.M., Tu, T-M. and Scott, K.P. (1983). Peroxidase-catalysed removal of phenols from coal-conversion waste waters, *Science* **221**, 259–260.
Marshall, W.G.A., Denault, L.J., Glenister, P.R. and Dower, J. (1982). Enzymes in brewing, *Brewers Digest* **57**, 14–22.
Park, Y.K., Martins, S.H. and Sato, H.H. (1983). Enzymatic removal of starch from sugar cane during sugar cane processing, *Process Biochemistry* **20**, 57–59.
Taylor, M.J., Olson, N.F. and Richardson, T. (1979). Coagulation of skim milk with immobilised proteases, *Process Biochemistry* **14**, 10–16.
Wandrey, C. and Wichmann, R. (1985). Coenzyme regeneration in membrane reactors, *Biotechnology Ser.* **5**, 177–208.
Ward, O.P. (1985). Hydrolytic enzymes, in *Comprehensive Biotechnology* Ed. Moo-Young, M., Vol. 3, pp. 819–836. Oxford, Pergamon Press.
Wiseman, A. (Ed.) (1985). *Handbook of Enzyme Biotechnology*, 2nd Edition. Chichester, Ellis Horwood.
Yokotsuka, T. (1985). Traditional fermented soybean foods, in *Comprehensive Biotechnology*. Ed. Moo-Young, M., Vol. 3, pp. 395–428. Oxford, Pergamon Press.

Chapter 8

Bergmeyer, H.U. (Ed.) (1986). *Methods of Enzymatic Analysis*, 3rd Edition. Weinheim, V.C.H.
Bowers, L.D. and Carr, P.W. (1978). Immobilised enzymes in analytical chemistry, *Advances in Biochemical Engineering* **15**, 89–129, New York, Springer-Verlag.
Caras, S. and Janata, J. (1980). Field effect transistor sensitive to penicillin, *Analytical Chemistry* **52**, 1935–1937.
Danielsson, B. (1985). Enzyme probes, in *Comprehensive Biotechnology*. Ed. Moo-Young, M., Vol. 4, pp. 395–422. Oxford, Pergamon Press.
Danielsson, B., Mattiasson, B., Karlsson, R. and Winqvist, F. (1979). The use of an enzyme thermistor in continuous measurements and enzyme reactor control, *Biotechnology and Bioengineering* **21**, 1749–1766.
Guibault, G. (1984). *Analytical Uses of Immobilised Enzymes*. New York, Marcel Dekker.
Hubble, J. (1986). The effect of operating conditions on the response of a differential enzyme thermistor, *Journal of Chemical Technology and Biotechnology*, **36**, 487–493.
Ichirose, N. (1986). Biosensors today and tomorrow, *Journal of Electronic Engineering* **23**, 80–87.
Lowe, C.R., Goldfinch, M.J. and Lias, R.J. (1983). Some novel biomedical sensors, *Biotech 83*, pp. 633–641. Northwood, Online Publications Ltd.
Mosbach, K. and Danielsson, B. (1981). Thermal bioanalyses in flow streams: enzyme thermistor devices, *Analytical Chemistry* **53**, 83A–94A.
Moss, S.D., Johnson, C.C. and Janata, J. (1978). Hydrogen, calcium, and potassium ion selective FET transducers: a preliminary report, *I.E.E. Transactions in Biomedical Engineering* BME **25**, 49–54.
Pedersen, H. and Horvath, C. (1981). Open tubular heterogeneous enzyme reactors in continuous-flow analysis, in *Applied Biochemistry and Bioengineering*. Eds. Wingard, L.B., Katchalski-Katzir, E. and Goldstein, L., Vol. 3, pp. 2–96. New York, Academic Press.
Plotkin, E.V., Higgins, I.J. and Hill, H.A.O. (1981). Methanol dehydrogenase bioelectrochemical fuel cell and alcohol detector, *Biotechnology Letters* **3**, 187–192.

Turner, A.P.F., Aston, W.J., Higgins, I.J., Bell, J.M., Colby, J., Davis G. and Hill, H.A.O. (1984). Carbon monoxide: acceptor oxidoreductase from *Pseudomonas thermocarboxydovorans* strain C2 and its use in a carbon monoxide sensor, *Analytica Chimica Acta* **163**, 161–174.
Weaver, J.C. and Burns, S.K. (1981). Potential impacts of physics and electronics on enzyme based analysis, in *Applied Biochemistry and Bioengineering*. Eds Wingard, L.B., Katchalski-Katzir, E. and Goldstein, L., Vol. 3, pp. 271–308. New York, Academic Press.
Williams, B.L. and Wilson, K. (1975). *Principles and Techniques of Practical Biochemistry*. London, Edward Arnold.
Williams, D.L., Doig, A.R. and Korosi, A. (1970). Electrochemical-enzymatic analysis of blood glucose and lactate, *Analytical Chemistry* **42**, 118–121.
Winqvist, F., Danielsson, B., Lurdstrom, I. and Mosbach, K. (1982). Use of hydrogen sensitive Pd-MOS materials in biochemical analysis, *Applied Biochemistry and Biotechnology* **7**, 135–139.
Yanchinsky, S. (1982). Biochips speed up chemical analysis, *New Scientist* **93**, 236.

Chapter 9

Clarke, P.H. (1980). Experiments in microbial evolution: new enzymes, new metabolic activities, *Proceedings of the Royal Society, London* **B207**, 385–404.
Danno, G. (1970). Studies on D-glucose-isomerizing enzyme from *Bacillus coagulans*, strain HN-68, *Agricultural and Biological Chemistry* **34**, 1805–1814.
Estell, D.A., Graycar, T.P. and Wells, J.A. (1985). Engineering an enzyme by site-directed mutagenesis to be resistant to chemical oxidation, *Journal of Biological Chemistry* **260**, 6518–6521.
Gacesa, P. and Venn, R.F. (1979). The preparation of stable enzyme-coenzyme complexes with endogenous catalytic activity, *Biochemical Journal*, **177**, 369–372.
Gait, M.J. (Ed.) (1984). *Oligonucleotide Synthesis—A Practical Approach*. Oxford, IRL Press.
Gronenborn, B. and Messing, J. (1978). Methylation of single-stranded DNA *in vitro* introduces a new restriction endonuclease cleavage site, *Nature*, **272**, 375–377.
Jacobsen, H., Klenow, H. and Overgaard-Hansen, K. (1974). The N-terminal amino-acid sequences of DNA polymerase I from *Escherichia coli* and of the large and small fragments obtained by proteolysis, *European Journal of Biochemistry* **45**, 623–629.
Kaiser, E.T. and Lawrence, D.S. (1984). Chemical mutation of enzyme active sites, *Science*, **226**, 505–511.
Smith, M. (1982). Site-directed mutagenesis, *Trends in Biochemical Sciences* **7**, 440–442.
Thomas, P.G., Russell, A.J. and Fersht, A.R. (1985). Tailoring the pH dependence of enzyme catalysis using engineering, *Nature* **318**, 375–376.

Chapter 10

Bender, M.L., D'Souza, V.T. and Lu, X. (1986). Miniature organic models of chymotrypsin based on α-, β- and γ-cyclodextrins, *Trends in Biotechnology* **4**, 132–135.
van Brunt, J. (1986). Protein architecture: designing from the ground up, *Biotechnology* **4**, 277–283.
Carrea, G. and Riva, S. (1986). Applications of cofactor dependent enzymes in organic synthesis, *Chimicaoggi* **3**, 17–21.
Hunkapiller, M.W. and Hood, L.E. (1983). Protein sequence analysis: automated microsequencing, *Science* **219**, 650–659.

King, J. (1986). Genetic analysis of protein folding pathways, *Biotechnology* **4**, 297–303.
Klibanov, A.M. (1986). Enzymes that work in organic solvents, *Chemtech*, 354–359.
Luisi, P.L. and Laane, C. (1986). Solubilization of enzymes in apolar solvents via reverse micelles, *Trends in Biotechnology* **4**, 153–160.
Sanger, F., Nicklen, S. and Coulson, A.R. (1977). DNA sequencing with chain-terminating inhibitors, *Proceedings of the National Academy of Sciences* **74**, 5463–5467.
Schmidt, E., Bossow, B., Wichmann, R. and Wandrey, C. (1986). The enzyme membrane reactor—an alternative approach for continuous operation with enzymes, *Chemistry and Industry* **35**, 71–77.
Suckling, C.J. (1984). Selectivity in synthesis-chemicals or enzymes, in *Enzyme Chemistry: Impact and Applications* Ed. Suckling, C.J., pp. 78–118. London, Chapman and Hall.
Ulmer, K.M. (1983). Protein engineering, *Science* **219**, 666–670.
Zaks, A. and Klibanov, A.M. (1984). Enzyme catalysis in organic media at 100°C, *Science* **224**, 1249–1251.

Appendix I

Webb, E.C. (Ed.) (1984). *Enzyme Nomenclature, 1984*. London, Academic Press.

Appendix II

Levenspiel, O. (1972). *Chemical Reaction Engineering*. New York, John Wiley & Sons.

Appendix III

General Reading
Cornish-Bowden, A. (1979). *Fundamentals of Enzyme Kinetics*. London, Butterworth.
Eisenthal, R. and Wharton, C. (1981). *Molecular Enzymology*. Glasgow, Blackie.

Index

Acetamide, 129
Acetate kinase, 150
Acetyl cholinesterase, 114
Acetyl phosphate, 150
Activated carbon, 98
Activated charcoal, 2
Activation energy, 46
  for enzyme decay, 78, 79
Active site modification, 136
Adenosine diphosphate, *see* ADP
Adenosine monophosphate, *see* AMP
Adenosine triphosphate, *see* ATP
ADP, 103, 147, 148, 150
Adsorption, 81
Affinity chromatography, 39, 70
  electrode, 118
Aflatoxins, 6
Albumin, determination of, 119
Alcohol dehydrogenase, 110
  oxidase, 110
Alginate, 26, 92
  lyase, 26
Alkaline phosphatase, 72, 111
Allergenicity, 6
Allergic reactions, 6, 7
Amidase, 129
Amines, aromatic, 94
L-Amino acid oxidase, 110, 112
Amino acid production, 98
Amino acids, 145
Aminoacylase, 4, 99
7-Amino cephalosporonic acid, 75
6-Aminopenicillanic acid (6APA), 74
Ammonia, determination of, 103
Ammonium, fumarate, 100
  ion electrodes, 119
  ion probe, 105
  sulphate fractionation, 35
AMP, 147, 148
Amperometric probes, 106, 108
Ampicillin, 75
  resistance, 25
Amylase, 10, 39, 93–98
Amyloglucosidase, 10, 95
Analysis of metabolites, 71
Analytical uses of enzymes, 71

Anaphylaxis, 69
Animal models for therapeutic studies, 69
Anti-leukaemic drugs, 69
Anti-oxidants, 79
Antibiotic resistance, 23, 25
Antibiotics, broad-spectrum, 74
  semi-synthetic, 73
Antibodies, 40, 66, 115
Antibodies, monoclonal, 40
Antigen, 66, 115
D-Arabinose, 130
Archaeobacteria, gene structure, 27
Arrhenius relationship, 52
  relationship for enzyme decay, 78
Artificial enzymes, 137
  organs, 67
Ascorbate oxidase, 145
L-Asparaginase, 68
  therapeutic use, 69–70
Aspartase, 4, 100
Aspartate, 69
Aspartate aminotransferase, 35
Aspartic acid, 100
Aspergillus, 122
Aspergillus niger, 71
Aspergillus oryzea, 92
ATP, 97, 147–150
ATP regeneration, 149–150
Attrition of immobilized enzyme particles, 88
Autolysis, 33
  of proteinases, 88
Autoradiography, 27
Axial mixing, 160

Bacillus amyloliquefaciens, 122, 136
Bacillus coagulans, 123
Bacillus licheniformis, 122
Bacillus subtilis, 21, 33
Backmixing, 159
Bacteria, 18
  Gram negative, 18, 19, 33, 74
  Gram positive, 18, 33
  growth cycle, 19
  pathogenic, 74
Ball mill, 42
Barbiturates, 68

Index

Barley, germination, 92
   malted, 15, 92
   malting, 90
Baseline stability, 115
Bating hides, 10
Beer, chill proofing, 10, 91
Benzene ring, 146
Benzoquinone, 109
Benzoyl tyrosine ethyl ester, 124
Benzoyl arginine ethyl ester, 141
Benzyme, 146
Biochips, 118
Bioelectronic sensor, 5
Biological oxygen demand (BOD), 96, 119
Biological washing powder, 6, 94
Biomass, 3
Biosensor, 5
Bis (phosphine) rhodium, 145
Bitter tasting peptides, 91
Blood, 69
Boltzmann constant, 83
Boltzmann distribution, 81–82
Boundary layer, 83, 84
Brewing, starch degradation, 92
Brewing industry, 92
Bromelain, 34, 91
8α-Bromo acetyl-10-methyisoalloxazine, 127
5-Bromo-4-chloro-3-indolyl
   β-D-galactopyranoside (X-gal), 133, 134
tert-Butanol, 144
Butyric acid, 144
Byproduct formation, 3

CM cellulose (CMC), 36–37
Carbamate kinase, 150
Carbamoyl phosphate, 150
Carbohydrate, degradation, 92
   wastes, 94
Carrageenan, 100
Casein, 91
   coagulation, 91
   isoelectric point, 91
Catabolite activator protein, 21–22
Catabolite repression, 21–22
Catalase, 115
Cell culture, animal, 7, 16
   plant, 7, 16
Cell membrane, 16
Cell structure, 4
Cell wall, 16
   bacterial, 18, 19
   enzyme degradation, 94
   fungal, 18, 19
   plant, 34
Cellobiase, 96
Cellulase, 20, 92–96
Cellulose, 92, 95
   degradation, 96

Centrifugation, 42
   differential, 33
   disk bowl, 42
   high speed, 42
   tubular bowl, 42
Cepalosporin C, 75
Cephaloridine, 75
Cephalosporins, determination of, 119
   semisynthetic, 75
Cetyl pyridinium chloride, 26
Cetyltrimethylammonium bromide, 142
Chaotropic ions, 40
Cheese, 96
   production, 91
   ripening, 90
Chelating agents, 34, 79
Chemical modification of enzymes, 79–80
Chemical mutagenesis, 130
Chitin, 18
Cholestenone, 142
Cholesterol, 71, 72, 142
   esterase, 72
   oxidase, 72, 115, 142
Chromatofocusing, 38
Chromatography, affinity, 39, 70
   adsorption, 98
   batch, 42
   gel permeation, 38
   HPLC, 40, 102
   hydroxyapatite, 125
   ion exchange, 36
Chymosin, 91
   cloned, 30
Chymotrypsin, 145
   mechanism, 147
Cibacron Blue F3G-A, 39, 40, 118
Clark electrode, (*see* oxygen electrode)
Cloning, urokinase gene, 70
Cloning vector, 23
Clupein, 124
Co-translational secretion, 16, 17
Coenzyme, 97, 101, 102, 126, 148
   retention in an enzyme reactor, 148
   immobilization, 148
Coenzyme regeneration, 5, 6, 101, 126, 137, 142, 147–149
   chemical, 148
   electrochemical, 148–149
Cofactor, 100, 126
   metal ions, 121
Cohn fractionation, 36
Computer, algorithm, 138
   generated models, 121
   modelling, 139
Conformational structure, 78
   changes on enzyme immobilization, 82
Conjugation, 68
Constitutive enzymes, 20

Index

Continuous stirred tank reactor, 159
Continuous monitoring, 115
Contraceptive pill, 75
Control of an enzyme reactor, 115
Corn starch, 97
Cos site, 27
Cosmids, 27
Cost effectiveness, ENFET's, 116
Creatinase, 112
Crown ethers, 147
18-Crown-6 ether, 147
Crystalline protein, 140
Crystallization, of proteins, 141
Curd, 90
Cyanogen bromide activated Sepharose, 74
Cyclic AMP, 22
Cyclodextrins, 145–147
Cyclomaltoheptaose (β-cyclodextrin), 145
Cyclomaltohexaose (α-cyclodextrin), 145, 146
Cylomaltooctaose (γ-cyclodextrin), 145
Cysteine, proteinase, 124
  active site, 127
Cystic fibrosis, 67
Cystosine deamination, 130–131
Cytochrome P450, 68

Deadzones, 159
DEAE cellulose, 36, 37, 81
Decay constant, 61, 77–78
Dehydrogenases, 126
Demethylation, 68
Denaturation, 34, 162
Deoxyribonucleic acid, *see* DNA
Derepression, 22
Desalting, 36
Desizing textiles, 10
Detergents, effects on enzyme stability, 78
Detoxification, 66, 68
Dextrose, 98
Diabetes, 118
Dialysis, 36, 42
Diamine oxidase, 110
Dibutyrin, 144
Dideoxynucleotide method, 140
Differential bridge, 114
Diffusion, bulk, 108
  effect on immobilized enzymes, 81
  internal, 108
Diffusivity, 84, 86
  through immobilization matrix, 86
Dilution rate, 57
Dimensionless groups, 84
Diosgenin, 76
Dioxane, 141, 144
Direct linear plot, 49, 164
Dissociating agents, 40
DNA, 23, 25, 27–28, 125, 130, 132, 138
  cDNA, 28–30

directed DNA polymerase, 28, 125
dependent RNA polymerase, 21–22
ligase, 133
packaging, 27
probe, 27, 134
RF DNA, 132–135
sequencing, 140
  single stranded, 130–135
Double reciprocal plot, 48
Downstream processing, 148
Disulphide bridge, 78

Economic considerations in the use of enzymes, 100
Edman degradation, 138–139
EDTA (ethylenediaminetetraacetic acid), 34, 79, 123
Effectiveness factor, 86
Electrode, 117
Electron acceptors, 109
Electrophoresis, 43
  sodium dodecyl sulphate, 43
Electrostatic interaction, 136
ELISA (enzyme linked immunosorbent assay), 71–73, 115
Encapsulation, 7
Endonucleases, 131
Endoplasmic reticulum, 16, 18
Enthalpy change, 114
Entrapment of enzymes, 81
Enzyme activity, 5
Enzyme assays, design of, 162
Enzyme Commission, nomenclature, 152
  numbers, 2, 152
Enzyme defects, 66
Enzyme electrode, 105–112, 118
  linear range, 108
  response time, 108
  wash time, 108
Enzyme encapsulation, for therapeutic uses, 66
Enzyme extraction, 32
  buffers, 34
Enzyme field effect transistors (ENFET's), 115
Enzyme kinetics, 45–55
  computer programs
Enzyme linked immunosorbent assay, 71–3, 115
Enzyme purification, 32–44
  heat treatment, 35
  large scale, 40
  organic solvents, 36
Enzyme sales, 11
Enzyme secretion, bacteria, 16
  eukaryotes, 16, 18
  fungi, 16
Enzyme specification, 43
Enzyme stability, 5, 77

Enzyme structure, prediction of, 137–138
  three dimensional, 139
Enzyme therapy, 65, 66
Enzyme thermistor, 109–115
Enzyme–coenzyme complexes, 126
Enzyme–drug conjugates, 115
Enzyme–electrode interactions direct, 116
Enzymes (see individual enzyme names)
  alphabetical listing, 152–158
  commercial preparations, 10
  covalent modification, 124
  operational stability, 4
  safety testing, 8
Enzymic modification of enzymes, 125
Equilibrium constant, 54
Erwinia caratovora, 69
Escherichia coli (E. coli), 21, 69, 125, 132
  β-galactosidase deficient, 131, 134
  immobilized, 100
  mismatch repair deficient, 134
Essential amino acids, 100
Esterification, 143
Eukaryote, gene structure, 27
European Economic Community, 11
Exons, 29
Exonuclease, 125, 131
Extracellular enzymes, 16, 19, 78
Extracorporeal shunt, 67, 70

F episome, 133
F-diagram, 159
Faraday constant, 107
Fermentation, 3, 76, 100
  ethanol production, 96
Ferrocene, 118
Fibrin, 70
Flavo enzymes, 127
  second order rate constants, 129
Flavo-papains, 127
  active site, 128
Flow injection analysis, 104
FMN oxidoreductase, 129
Food additives, 7
Formaldehyde, 117
Formate dehydrogenase, 101, 149
Fractional conversion, 57
French pressure cell, 33
Fructose, 97
Fruit juice, 94
L-Fucose isomerase, 130
Fuel cell, 116–117
Fungi, 16

Galactose, 96
β-Galactosidase, 4, 21, 41, 72, 97, 131–133
Galactoside permease, 21
Gas constant, 107
Gelatin, 91

Gene, 134
  cloning, 22–23
  expression, 25
  structure, 22
Genetic, defects, 66
  information, 130
  manipulation, 15, 22–23
Glucanases, 93
Glucans, 18
Glucoamylase (amyloglucosidase), 4, 97–98
Glucose, 71, 96, 97
  cyclic polymers of, 145
  determination of, 103, 105, 119
Glucose 6 phosphate, 150
Glucose 6 phosphate dehydrogenase, 39, 103
Glucose isomerase, 4, 11, 59, 81, 97, 122
  immobilized, 100
  substrate specificity, 123
Glucose oxidase, 71, 72, 105, 110, 115, 118
α1–4 Glucosidase, 67
Glucosidase, 112
β-Glucuronidase, 35
Glutamate dehydrogenase, 103
Gluten, 92
Glycanohydrolases, 125
Glycohydrolases, 15
Glycoprotein, 69, 125
  sialic acid terminated, 66
  mannose terminated, 71
Glycosaminoglycans, 71
Glycosidases, 11
Glycosylation, 18, 30
Golgi apparatus, 16, 18
Good Manufacturing Practice, 7
Grass fermentation, 92
Gratuitous inducers, 20

Haemodialysis, 67
Haemolytic streptococci, 70
Haldane relationship, 55
Half life, 4, 78, 143
  aminoacylase, 99
  enzyme in serum, 66, 69
  glucose isomerase, 100
Halophiles, 79
Health & Safety at Work Act, 7
Heat, capacity, 113
  shock, 100
  stability, 121
  transfer, 42
α-Helices, 145
n-Heptanol, 143, 144
Hexokinase, 103, 115, 150
High fructose corn syrup (HFCS), 97–100
Histidine ammonia lyase, 4
HMSO, 7
Homogenisation, 33, 42
Hyaluronidase, 35, 70, 71

Index

Hybridization, 26
Hydrogen bonding, 78
Hydrolases, 143
Hydrolysis, 144
Hydrolytic enzymes, 72, 147
Hydrophobic, compounds, 141
   interactions, 78
Hydrophobicity, 124
Hydroxylation, 68
20-β-Hydroxy steroid dehydrogenase, 142

Imidazole ring, 147
Immobilization, coenzymes, 126, 148
Immobilization effects on enzyme activity, 81
   conformational, 81
   diffusional, 81
   partitioning, 81
Immobilization effects on enzyme stability, 88
Immobilization of enzymes, 74, 77–89, 124
   by adsorbtion, 81
   by covalent bonding, 81
   by encapsulation, 81
   by entrapment, 81
   using ultrafiltration membranes, 81
Immobilization of mycelial pellets, 94
Immobilized antibody, 115
Immobilized cell, 3, 4, 97, 100, 119
Immobilized enzymes, 2–7, 56–58, 97, 99, 103, 109, 148
   economics, 5
   product consumption, 12
   reactor, 104
Immobilized ligand, 39
Immobilized NAD(H), 150
Immune response, 6
Immunity, acquired, 66
Immunoadsorbents, 40
Immunological, reaction, 70
   response, 66
Inactivation of enzymes, 4
Indirect enzyme linked assays, 72
Indole, 100
Inducible enzymes, 20, 21
Industrial, chemical production, 11
   enzymes, 11, 16, 30
   waste, 95
Industry, brewing, 15
   enzyme, 137
   food, 7, 11, 15
   meat, 15
   pharmaceutical, 7
Inhibition, competitive, 50–51, 61
   product, 50
   reactant, 51
   uncompetitive, 50, 60
Inner membrane, 18
Insertional inactivation, 25, 134
Integrated circuit, 115

International Union of Biochemistry, 2, 4, 152
Intracellular enzymes, 16, 33, 42, 78
Intravenous administration of enzymes, 66
Introns, 27, 29
Inulin, 33
Inulinase, 33
Ion, exchange membrane, 117
   selective electrodes, 105, 115
   selective field effect transistors (ISFET's), 115–116
Ionic bonds, 78
Ionic strength, 34
   effect on adsorbed enzymes, 81
ISFET (*see* ion selective field effect transistor)
Isoalloxazines, 127
Isoelectric focusing, 43
Isoelectric point, 69
Isoenzyme, 66
Isomorphous crystals, 139

Jack bean meal, 119

20-Keto steroids, 142
Kidney failure, 67
Kinases, 126
Kinetic constants determination of, 48
Klebsiella pneumoniae (*Klebsiella aerogenes*), 26, 33, 130
Klenow enzyme, 125, 134
Kluyveromyces, 30
Kluyveromyces marxianus, 33
Kluyveromyces fragilis, 97
Kluyveromyces lactis, 97
K_m (Michaelis constant), 69, 141, 162–164
K_m, alteration, 134
   effect on linear range of a sensor, 109
K_m, importance of low values for therapeutic enzymes, 69
Koji, 10

Lac operon, 21
Lac Z gene, 133
Lac-repressor protein, 22
Lactamase (penicillinase), 74
Lactate, estimation of, 105
Lactate dehydrogenase, 2, 111, 115, 126
Lactic acid, 91
Lactoperoxidase, 144
Lactose, 91, 94
   hydrolysis, 96
Laminar flow, 85
Legislation, 6, 7, 43
Leukaemia, acute, 68
Ligation, 23, 133, 134
Light emitting diode (LED), 118
Lignin, 96, 144
Ligninase, 144
Limit dextrin, 98

Lipase, 91, 94, 143–144
Liposomes, 67
Liquid/liquid extraction, 42
Liver artificial, 67
Lysis, 33, 134
Lysosomal storage diseases, 67

M13 phage, 131–138
  life cycle, 132
  selection of recombinants, 133, 134
MAFF, 7
Mass transfer, 42, 83–89, 105, 108, 109, 116, 118
  external, 81–83
  internal, 81–86
  resistance, 84
Meat, flavour enhancement, 90
  tenderization, 90
Membrane bound enzymes, 81
Membrane properties, sensors, 116
Metabolic defects, 66
Metal ion cofactor, 123
Metal ions, substitution of bound, 122
Metalloproteinases, 34
Metastable complex, 46
Methanol, 117
Methanol dehydrogenase, 117
Methionine, substitution, 136
Michaelis constant (see K_m)
Michaelis Menten equation, 47, 162, 165
Michaelis Menten equation, integrated form, 55, 164
Microbial, degradation of wastes, 94
  evolution, 129
  rennet, 10, 30
Microbiological activity, 6
Microcalorimetry, 109
Microencapsulation, 81
  therapeutic enzymes, 70
  urease, 68
Microorganisms, 129
  eukaryote, 33
  for enzyme production, 9
  for use in the food industry, 9
  pathogenic, 22
Microsomal-bound enzymes, 68
Milk, 90
Milk products, 91
Milling, cellulose processing, 96
  corn, 97
Molar, enthalpies, 115
  extinction coefficient, 103
Molecular graphics, 139
Morteirella vinaceae, 94
Mucor miehei, 91
Multi-enzyme, synthesis, 5, 6
  systems, 104

Mutagenesis
  non-specific, 129
  oligonucleotide, 132, 134, 135
  site-specific, 132, 138
Mutagenic oligonucleotides, 134
Mutation, 134
Mycotoxins, 6
Myocardial infarction, 66, 70
Myoglobin, 145

NAD, 5, 39, 101, 103, 126, 148, 149
NAD(P), 102, 147, 148, 150
NAD(P)H, 68, 147, 148
NADH, 103, 149
NADH regeneration, 142
Neoplasm, 69
  control of, 68
Nick translation, 28
Nicotinamide adenine dinucleotide (*see* NAD)
  phosphate (*see* NAD(P))
  reduced (*see* NADH)
  phosphate reduced (*see* NAD(P)H)
Nitrate, estimation of, 105
NMR, 140
Non-enzymic proteins, 145
Norcardia, 142
Nuclear magnetic resonance (*see* NMR)
Nuclear Overhauser spectroscopy, two dimensional, 140
Nucleophile, 143
Nylon, 105

Old yellow enzyme, 129
Oligonucleotide, 132
  mutagenesis, 132, 134, 135
Oligosaccharides, 125
Open tubular reactors, for use in analysis, 105
Operational stability, 142
  enzyme based processes, 78–79
Optical resolution, 144
Opto-electronic sensors, 118–119
Organic, acid, 100
  phase, 142
  polymers, 79
  solvent, 78, 141–144
Organic solvents, use of enzymes in, 141–144
Organometal complexes, 145
Outer membrane, 18
Oxidoreductases (*see* dehydrogenases), 126–127
Oxygen electrode, 108
  half cell reactions, 108
Oxygen probe, 105–119

Packed bed reactor, 84, 86, 97, 104
Pancreas, 143
Pancreatic, enzymes, 67
  lipase, 15
Papain, 10, 15, 34, 91, 124, 127

Index

Patent protection, 125
Pectinase, 10–11
Penicillin, amidase, 74
  determination of, 119
  G (Benzyl penicillin), 74
  V (Phenoxymethylpenicillin), 74
Penicillinase (lactamase), 74, 112, 116
Pentosanases, 93
Pepsin, 10, 91
Peptide hormones, 71
Peptide structure, 145
Peptidoglycan, 18
Perfect mixing, 160
Periplasmic space, 18
Peroxidase, 72, 144
  estimation of, 114
  horseradish, 94, 144
pH effect on, adsorbed enzymes, 81
  enzyme activity, 52
  enzyme stability, 77
  microenvironment, 81
  purification, 34–35
  storage stability, 78
pH measurement as a basis for analysis, 107
pH microenvironment, 83
pH optimum, 124
  glucose isomerization, 98
  immobilized enzyme, 82
pH probe, 107
  half cell reactions, 106
pH profiles, 136
pH sensitive dyes, 119
Phage λ, 27
  genome size, 27
Phage, M13 (*see* M13 phage)
Pharmaceutical applications of enzymes, 73
Phenazine ethanosulphate, 117
Phenol degradation, 94
Phenols, 94
Phenyl isothiocyanate, 139
Phenylacetamide, 130
D-Phenylglycine, 100
Phenylketonuria, 66
Phosphite-triester chemistry, 134
Phosphotriester chemistry, 134
Photocell, 118
pK_a, 124, 136
Plant extracts, 30
Plaques, 134
Plasma, 102
Plasmids, 23, 26
Plasmin, 70
Plasminogen, 70
Platinum cathode, 108
β-Pleated sheet, 145
Plug flow, 161
  reactor, 58–59, 159
Polyamino polystyrene resin, 2

Polypeptide, primary sequence of, 138
Polyphasic activity decay, 88
Polyphenol oxidase, 34
Polysaccharides, 38, 93
Polytetrafluoroethylene (PTFE), 108
Pompe's disease, 67
Porosity, 86
Post translational modifications, 18, 30, 138
Potentiometric probes, 106
Power dissipation, 114
Primary protein sequence, 78
Process costs, effect of stability, 77
Process economics, 6, 137
Proprionamide, 129
Protamine sulphate, 36
Protein, engineering, 79
  folding, 138
  sequencing, 138
Proteinase, cysteine, 15
Proteinases, 6, 10–11, 21, 33, 88, 90–94, 124–125, 144
Proteolysis, 91, 145
Proteolytic enzymes (*see* proteinases)
Proteolytic modification of enzymes, 125
Pseudomonas aeruginosa, 130
PTFE (*see* polytetrafluoroethylene)
Pullulanase, 19, 33, 98

Racemic mixture, 98
Radiation, 130
Radioimmunoassay, 72
Raffinose, 94
Reactant concentration, effects on rate of reaction, 47
  effects on stability, 78
Reaction rate, 45
Reactor design, 45, 55–64
  choice of reactor type enzyme requirement, 59
  enzyme inactivation, 63
  inhibition effects, 60
  kinetics of batch reactors, 55
  kinetics of continuous processes (CSTR), 56
  kinetics of continuous processes (PFR), 58
  performance equations, 62
  stability effects, 61
Recombinant DNA, techniques (*see* genetic manipulation)
Redox dyes, 72, 117
Renal failure, 67
Renaturation, 79
Rennet, 10, 11
Replication, 23
Residence time, 58
  distribution, 159
Restriction endonuclease Type II, 23
Restriction endonucleases, 25
Reverse micelles, 141, 142

Reverse transcriptase (*see* RNA directed DNA polymerase)
Reverse transcription, 27, 29
Reynolds number, 85
Ribonuclease H, 28
Ribonucleic acid (RNA), 27
Ribose moiety, 148
Ribosomes, 16
RNA directed DNA polymerase, 27
RNA polymerase (*see* DNA-directed RNA polymerase)
RNA probe, 27
Ruthenium transfer catalyst, 145

Schardinger dextrans, 145
Schmidt number, 85
Seaweed, 94
Secondary metabolites, 6
Segmented flow analysis, 105
Selwyn's test for enzyme inactivation, 163
Semiconductor materials, 109
Sensitivity, of sensor, 109–114
Sensors, enzyme based, 102–120
Serratia marcescens, 69
Serum, guinea pig, 69
Sherwood number, 84–85
Shot-gun cloning, 23, 27
Shuttle vectors, 30
Signal peptide, 16
Silage production, 92
Silver anode, 108
Single cell protein, 100
Site directed mutagenesis, 140
Solid phase synthesis, 134
Solvents, non-aqueous, 100
  organic, 137, 141, 143
Sophorose, 20
Sources of enzymes, animals, 15, 30
  microorganisms, 9, 16–31
  plants, 15
Soy protein hydrolysates, 92
Spatial structure, 140
Spectrophotometer, 163
Spectrophotometric, assay, 72
  techniques, 119
Spleen, 67
Stability, enzyme, 4, 5, 77–80
  of modified enzymes, 136
  of reactor performance, 61, 78
  of reactor performance, 78
Stability assessment, storage and operation, 78
  immobilized enzymes, 88
Stabilization of an enzyme for storage, 79
Starch, 95
  acid hydrolysis, 97
  conversion, 97
Steady state assumption, 46
Steam explosion, cellulose processing, 96

Stereospecificity, 129
Steric hinderance, 81
Steroids, 75
Stigmasterol, 76
Stirred tank reactor, 84
Streptokinase, 70
Sub-cellular organelles, 33
Substrate specificity, 124
Subtilisin BPN, 136
Subtilisin Carlsberg, 124, 125
Sugar beet molasses, 94
Sugar refining, 94
Super conducting magnets, 140
Superficial liquid velocity, 84
Sweetners, 97
Synchrotron X-ray source, 139
Synthetic enzymes, 144

Temperature, effect on activity, 4, 52
  effects on enzyme stability, 77
Tenderizing agents, 91
Tetracycline resistance, 25
Tetranitromethane, 124
Texture modification, 91
Therapeutic index, 69
Thermal enzyme-linked immunosorbent assay (TELISA), 115
Thermal, inactivation, 143
  stability of glucose isomerase, 97
Thermistor, 109
  response, 113
Thermophiles, 79
Thermostable enzymes, 122
Thiele modulus, 86
  plot, 87
Thiogalactosidase transacetylase, 21
Three dimensional, 141
  spatial proton map, 141
  structure, 138
p-Toluene sulphonyl-L-arginine methyl ester, 124
Tortuosity, 86
Toxicity, 6
Transcription, 21
Transducer, 104, 109
  bound enzymes, 105
Transesterification, 143, 144
Transformants, 26
Transformation, 23, 133
Tributyrin, 143, 144
Trichoderma reesei, 96
Trypsin, 10, 15, 84, 115, 125
  effect of organic solvents, 141
Trypsinogen, 125
L-Tryptophan 100
Tumour cells (*see* neoplasms),
Turbulent flow, 85

Index

Type II glycogen storage disease (Pompe's disease), 67

Ultra violet light, 130
Ultra-low water system, 142
Ultrafiltration, 36, 42, 148
  membrane, 101
Uracil, 130
Urea, 67
  determination of, 119
Urease, 2, 34, 67, 105, 111, 115
  use in ELISA, 73
Uric acid oxidase, 111
Uricase, 115
Urokinase, 70

Valeramide, 130

Vector, 134
Vesicles, 16

Waste treatment, 94
Water, essential, 143
  in oil microemulsions, 141
  pollution, 119
Western blotting, 43
Whey, 96
Wild type enzymes, 136

X-gal, 131, 132
X-ray diffraction, 139–141
Xylose isomerase, 97, 122
  substrate specificity, 123

Yeast, 30, 143